Global environmental issues

Worldwide summits, government-sponsored studies, and the emergence of the Greens as a force in Western politics highlight a concern for global environment issues which embraces an increasingly broad spectrum of society. The complexity of the problems and the need to meet the surge of interest in them has spawned a literature polarized between the highly technical and very general with little in between. *Global Environmental Issues* was written to fit the middle ground. Professor Kemp discusses the problems – the greenhouse effect, ozone depletion, acid rain, nuclear winter, and drought – by way of the common thread of climatology which serves to underline the global dimension of such issues. He also emphasizes the deteriorating relationship between society and its environment, an element shared by all of the problems. The book does not offer any magical solutions, but rather warns of the dangers of neglecting the climatic factor in long-term plans for the protection of the environment.

The multi-disciplinary, non-technical nature of the book should appeal to students of geography and environmental studies and other disciplines where the environmental approach is emphasized. It is also suitable for the non-academic reader and those whose work requires an understanding of global environmental issues.

David D. Kemp is Associate Professor in the Department of Geography at Lakehead University, Thunder Bay, Ontario.

Global environmental issues
A climatological approach

David D. Kemp

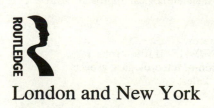

London and New York

First published 1990
by Routledge
11 New Fetter Lane, London EC4P 4EE

Simultaneously published in the USA and Canada
by Routledge
a division of Routledge, Chapman and Hall, Inc.
29 West 35th Street, New York, NY 10001

Typeset by Leaper & Gard Ltd, Bristol, England
Printed and bound in Great Britain by
Mackays of Chatham PLC, Chatham, Kent

British Library Cataloguing in Publication Data

Kemp, David, D. *1943–*
    Global environmental issues: a climatological approach.
    1. Environment.   Effects of climate.
    I. Title
    333.7

ISBN 0-415-01108-6
ISBN 0-415-01109-4 pbk

Library of Congress Cataloging in Publication Data

Kemp, David D., 1943–
    Global environmental issues : a climatological approach / David D.
    Kemp.
        p.   cm.
    Includes bibliographical references.
    ISBN 0-415-01108-6. — ISBN 0-415-01109-4 (pbk.)
    1. Pollution.   2. Air—Pollution—Meteorological aspects.
    3. Pollution—Social aspects.   I. Title.
    TD177.K46   1990
    363.73'92—dc20   89-70244
                                                                 CIP

To Susan, Heather, Qais, Qarma,
Beisan, Alisdair, Colin, Mairi,
Eilidh, Euan, Neil, Ross, Andrew,
and all of their generation whose
future requires that we find
solutions to our current
environmental problems

Contents

List of tables

List of figures

Acknowledgements

Although the author's name is the only one to appear on the title page, any book is the culmination of the combined efforts of a number of individuals. This one is no exception.

Mary Ann Kernan of Routledge deserves special thanks for her interest in the original concept and for guiding a first-time author painlessly through the pitfalls of academic publication. Gwen Henry provided her skills on the word processor to type the numerous drafts of the book, and coped cheerfully with what must have seemed to be an endless series of revisions. Her help is very much appreciated. Thanks are also due to Geoff McBoyle, Professor of Geography at the University of Waterloo, for his helpful comments on the penultimate draft. I am particularly grateful to my wife and family for their patience, encouragement, and understanding during the writing of the book.

I would also like to thank the following for allowing me to reproduce copyright material in the book:

Figure 3.3 Reprinted with permission from *Forum Africa* 1985, Canadian International Development Agency.

Figure 3.6 This first appeared in *New Scientist* magazine, London, the weekly review of science and technology, in Cross, M. (1985) 'Africa's drought may last many years'. *New Scientist*, 105 (1439): 9.

Figure 3.10 Reprinted with permission from Bryson, R.A. and Murray, T.J. (1977) *Climates of Hunger*, Madison, Wisconsin: University of Wisconsin Press.

Figures 3.11 and 3.12 Reprinted with permission from Rasmusson, E.M. and Hall, J.M. (1983) 'El Niño', *Weatherwise* 36(4): 166–75, Washington DC: Heldref Publications.

Figures 7.2 and 7.3 Reprinted with permission from SCOPE 29: Bolin, B., Boos, B.R., Jagar, J., and Warrick, R.A. (1986) *The Greenhouse Effect, Climatic Change and Ecosystems*, New York: John Wiley & Sons.

Figure 7.4 Reprinted with permission from Schneider, S.H. and Mass, C. (1975) 'Volcanic dust, sunspots and temperature trends', *Science* 190: 21 November, 741–6. Copyright 1975 by the AAAS.

Figure 8.3b Reprinted with permission from Schneider, S.H. and Thompson, S.L. (1988) 'Simulating the climatic effects of nuclear war', *Nature* 333: 221–7. Copyright 1988 Macmillan Magazines Ltd.

Preface

The study of global environmental problems has produced an abundance of published material, ranging from technical reports to basic articles prepared for popular consumption. The bulk of the material has accumulated at either end of that spectrum, and the area in between has been less well served. This book is designed to fit into the gap in that middle area.

It is an introductory text dealing with modern global environmental issues – such as the greenhouse effect, acid rain, ozone depletion, nuclear winter, atmospheric turbidity, and drought – all of which include a climatological element in their make-up. It emphasizes the relationships between society and environment, which are central to all of these issues, and identifies elements shared by them. This allows the recognition of common causes and the possibility of common solutions.

The book is multi-disciplinary in its approach, suitable for introductory courses in geography and environmental studies programmes, but is also appropriate for courses in other disciplines where the environmental approach is identified. As an introductory text it is designed to provide a broad understanding of the issues for those who wish only that, yet with sufficient detail to provide a base for those who may wish to develop a greater knowledge of one or more of the topics.

Questions on the working of the atmosphere are common in any discussion of global environmental issues. This book therefore includes a chapter which gives some consideration to the atmosphere but does not attempt to provide a comprehensive summary of that medium. Rather, it considers specific elements such as the composition of the atmosphere, its general circulation, and the energy budget of the earth/atmosphere system. All of these are involved in the creation or intensification of current global environmental problems. In this chapter, a conscious attempt was made to reference a variety of well-established introductory climatology texts which could be consulted by students who have little previous experience of atmospheric studies.

There is an abundance of literature on the topics covered in this book. A number of papers are already considered classic or are destined to become so. In contrast, some of the less academic material appearing in newspapers and popular magazines may be considered suspect by some specialists. In the rapidly expanding investigation of global environmental problems, however, the first indications of important discoveries or new control initiatives are often to be found in the popular press. At the introductory level, such material can be used quite effectively to lead students into an appreciation of the more complex papers which appear in technical journals. The references quoted in this book therefore include articles prepared for popular consumption as well as those written for and by the academic community.

Each chapter includes suggestions for further study which encourage the student to consider some of the ideas presented in the text. The problem of scale is a common theme in many of these sections. All of the environmental concerns are characterized by their global extent, therefore the further-study sections reinforce that by including ideas for debate and discussion which direct the reader to examine the global implications of the problems. In contrast, other questions ask the students to consider the local impacts of the issues by identifying activities in their own communities which contribute to global environmental problems or are affected by them.

If the experts are right, the full effects of current global environmental problems will be felt within the lifetime of the students using this book. I hope that the following chapters will improve their understanding of the problems, and allow them to recognize the importance of facing up to the issues before it becomes too late.

David D. Kemp
Thunder Bay
May 1989

Chapter one

Setting the scene

Writing in 1972, the climatologist Wilfrid Bach expressed concern that public interest in environmental problems had peaked and was already waning. His concern appeared justified as the environmental movement declined in the remaining years of the decade – pushed out of the limelight in part by growing fears of the impact of the energy crisis. In the 1980s, however, there has been a remarkable resurgence of interest in environmental issues, particularly those involving the atmospheric environment. The interest is broad, embracing all levels of society, and holding the attention of the general public plus a wide spectrum of academic, government, and public-interest groups.

Most of the issues are not entirely new. Some – such as acid rain, the enhanced greenhouse effect, atmospheric turbidity, and ozone depletion – have their immediate roots in the environmental concerns of the 1960s, although the first two had been recognized as potential problems in the nineteenth century. Drought and famine are problems of even longer standing. Contrasting with all of these is nuclear winter, which remains an entirely theoretical problem at present, but potentially no less deadly because of that.

Diverse as these issues are, they have a number of features in common. They are, for example, global or at least hemispheric in magnitude – large-scale compared to the local or regional environmental problems of earlier years. All involve human interference in the atmosphere component of the earth/atmosphere system, and this is perhaps the most important element they share. They reflect society's ever-increasing ability to disrupt environmental systems on a large scale.

Scientists have long recognized the ability of natural processes to change the atmosphere, both chemically and physically, but have always assumed that in a matter of years – decades at most – it would return to some normal state as a result of built-in checks and balances. Until relatively recently, society's contribution to atmospheric change was small enough that it was assumed these checks and balances could also deal with disruption caused by human activities. As the global problems which have emerged over the past several years become better understood, however, it is apparent that such an assumption may no longer be tenable.

The impact of society on the environment

Society's ability to cause major changes in the atmosphere, and other elements of the environment, is a recent phenomenon, strongly influenced by demography and technological development. Primitive peoples, for example, being few in number and operating at low energy levels with only basic tools, did very little to alter their environment. The characteristically low natural growth rates of the hunting and gathering groups who inhabited the earth in prehistoric times ensured that populations remained small. This, combined with their nomadic life-style and the absence of any mechanism other than human muscle by which they could utilize the energy available to them, limited their impact on the environment. In truth, they were almost entirely dominated by it. When it was benign, survival was assured. When it was malevolent, survival was threatened. Population totals changed little for thousands of years, but slowly, and in only a few areas at first, the dominance of the environment began to be challenged. Central to that challenge was the development of technology which allowed the more efficient use of energy (see Table 1.1). It was the ability to concentrate and then expend larger and larger amounts of energy which made the earth's human population uniquely able to alter the environment. The ever-growing demand for energy to maintain that ability is at the root of many modern environmental problems (Biswas 1974).

The level of human intervention in the environment increased only slowly over thousands of years, punctuated by significant events which helped to accelerate the process. Agriculture replaced hunting and gathering in some areas, methods for converting the energy in wind and falling water were discovered, and coal became the first of the fossil fuels to be used in any quantity. As late as the mid-eighteenth century, however, the environmental impact of human activities seldom extended beyond the local or regional level. A global impact only became possible with the major developments in technology and the population increase which accompanied the so-called Industrial Revolution. Since then – with the introduction of such devices as the steam engine, the electric generator, and the internal combustion engine – energy

Table 1.1 Energy use, technological development, and the environment

Time	Daily per capita energy consumption	Main sources	Use	Environmental impact
1,000,000 BC	2,000 kcals	Food, human muscle	Daily life	Minimal
100,000 BC	4–5,000 kcals	Food, fire, simple tools	Heating, cooking, hunting	Local and short term, mainly vegetation destruction and reduction of animal population
5,000 BC	12,000 kcals	Animals, agricultural produce	Transportation, cultivation, construction	Local and longer term, mainly in agricultural hearths; natural vegetation replaced by cultivated crops; aquatic environment altered; beginnings of soil degradation
AD 1400	26,000 kcals	Wind, water, coal, windmills, water wheels	Mechanical operations, pumping water, sawmilling, grinding grain, transportation	Local and longer term or permanent; natural vegetation removed; urban air pollution already common
AD 1800	50,000 kcals	Coal, steam engine	Mechanical operations, industrial processes, transportation	Local and regional; permanent major landscape changes begin; air and water pollution common in industrial areas
AD 1980	300,000 kcals	Fossil fuels, nuclear energy, internal combustion engine, electricity	Mechanical operations, industrial processes, transportation, social and cultural development	Local, regional, and global; permanent and perhaps irreversible air, water, and soil deterioration on global scale; acid rain; enhanced greenhouse effect; ozone depletion; increased atmospheric turbidity

Source: Compiled from data in Biswas (1974), Kleinbach and Salvagin (1986)

consumption has increased six-fold, and world population is now five times greater than it was in 1800. The exact relationship between population growth and technology remains a matter of controversy, but there can be no denying that, in combination, these two elements were responsible for the increasingly rapid environmental change which began in the mid-eighteenth century. At present, change is often equated with deterioration, but in the mid-eighteenth century technological advancement promised such a degree of mastery over the environment that it seemed such problems as famine and disease, which had plagued mankind for centuries, would be overcome, and the quality of life of the world's rapidly expanding population would be infinitely improved. That promise was fulfilled to some extent, but, paradoxically, the same technology which had solved some of the old problems also exacerbated others, and ultimately created new ones.

The response of the environment to human interference

The detrimental effects of the changes were not appreciated at the time, for their full effects were dampened to some extent, even as they developed, by the inherent stability of the environment. Changes in one element in the system which tended to produce instability were countered by changes in other elements which attempted to restore balance. This tendency for the components of the environment to achieve some degree of balance has long been recognized by geographers, and referred to as environmental equilibrium. The balance is never complete, however. Rather, it involves a continuing series of mutual adjustments among the elements which make up the environment. The rate, nature, and extent of the adjustments required will vary with the amount of disequilibrium introduced into the system, but in every environment there will be periods when relative stability can be maintained with only minor adjustments. At other times, the equilibrium is so disturbed that stability is lost, and major responses are required to restore the balance. Many environmentalists view the present environmental deterioration as the result of human interference in the system – to a level which has pushed the stabilizing mechanisms to their limits, and perhaps beyond.

Running contrary to this is the much less pessimistic point of view expressed by James Lovelock and his various collaborators in the Gaia hypothesis (Lovelock 1972, Lovelock and Margulis 1973). First developed in 1972, and named after an ancient Greek earth-goddess, the hypothesis views the earth as a single organism in which the individual elements exist together in a symbiotic relationship. Internal control mechanisms maintain an appropriate level of stability. Thus, it has much in common with the concept of environmental equilibrium. It goes further, however, in presenting the biocentric view that the living component of the environment is capable of working actively to provide and retain optimum conditions for the survival of life. This is one of the more controversial aspects of Gaia, flying in the face of majority scientific opinion, which sees life responding to environmental conditions rather than initiating them, and inviting some interesting and possibly dangerous corollaries. It would seem to follow, for example, that existing environmental problems which threaten life – ozone depletion, for example – are transitory, and will eventually be brought under control again by the environment itself. To accept that would be to accept the efficacy of regulatory systems which are as yet unproven, particularly in their ability to deal with large-scale human interference. Such acceptance would be irresponsible, and has been referred to by Schneider (1986) as 'environmental brinksmanship'.

Lovelock himself has allowed that Gaia's regulatory mechanisms may well have been weakened by human activity (Lovelock and Epton 1975, Lovelock 1986). Systems cope with change most effectively when they have a number of options by which they can take appropriate action, and this was considered one of the main strengths of Gaia. It is possible, however, that the earth's growing population has created so much stress in the environment, that the options are much reduced, and the regulatory mechanisms may no longer be able to nullify the threats to balance in the system. This reduction in the variety of responses available to Gaia may even have cumulative effects which could threaten the survival of the human species. Although the idea of the earth as a living organism is a basic concept in Gaia, the hypothesis is not anthropocentric. Humans are simply one of the many forms of life in the biosphere, and whatever happens life will continue to exist, but it may not be human life. It seems unlikely that problems such as acid rain and the enhanced greenhouse effect would threaten human existence, even allowing for their cumulative effects. However, ozone depletion, with its ability to cause genetic change, and nuclear winter, in association with the other effects of nuclear war, pose a serious threat to life.

The need for co-existence between people and the other elements of the environment, now being advocated by Lovelock as a result of his research into Gaia, has been accepted by several generations of geographers, but historically society has tended to view itself as being in conflict with the environment. Many primitive groups may have enjoyed a considerable rapport with their environment, but, for the most part, the relationship was an antagonistic one (Murphey 1973). The environment was viewed as hostile, and successful growth or development depended upon fighting it – and winning. In the beginning, human inputs were relatively minor, and the results of victory could easily be accommodated in the system. Gradually, through technological advancement and, sometimes, sheer weight of numbers, society became increasingly able to challenge its environment and eventually to dominate it. Natural vegetation was replaced by cultivated crops, rivers were dammed or diverted, natural resources were dug from the earth in such quantity that people began to rival geomorphological processes as agents of landscape change, and, to meet the need for shelter, nature was replaced with the built environment created by urbanization.

The environment can no longer be considered predominantly natural in most of Europe and North America. Technological innovations since the mid-eighteenth century have ensured that. In the less-developed nations of Asia, Africa, and South America – where the impact of technology is less strong – pressure from large and rapidly growing populations has placed an obviously human imprint on the landscape. Extensive as it is, however, dominance is far from complete. Even after 200 years of technological development and the exponential growth of population in recent years, some geographical regions continue to resist human domination. The frozen reaches of the Arctic and Antarctic are not unaffected by human activity, but they are largely unpopulated, while habitation in the world's hot deserts is in all senses marginal.

Human activities and the atmospheric environment

Certain elements in the environment remain untamed, uncontrollable, and imperfectly understood, and nowhere is this more evident than in the realm of weather and climate. Neither nineteenth- nor twentieth-century technology could prevent cyclones from devastating the shores of the Bay of Bengal, or hurricanes from laying waste the Caribbean. The Sahelian drought spread uncontrollably even as the first astronauts were landing on the moon. The developed nations, where society's dominance of the environment is furthest advanced, continue to suffer the depredation of tornadoes and blizzards as well as the effects of less spectacular weather events such as frost, drought, heatwaves, and electrical storms, which even today are difficult to forecast. No one is immune.

It is scarcely surprising, therefore, that weather and climate are universal topics of interest at all levels of society. In most cases, concern centres on the impact of short-lived local weather events on individuals and their activities. It is very much one-sided, normally ignoring the potential that mankind has to alter its atmospheric environment. There is a growing appreciation, however, that the nature and extent of the climatological component in many current global issues is strongly influenced by human interference in the earth/atmosphere system. The evolution of this awareness has been traced by William Kellogg (1987) who points out that as long ago as the late-nineteenth century, the first tentative links between fossil fuels, atmospheric carbon dioxide, and world climate had been explored. The results failed to elicit much interest in the scientific community, however, and remained generally unknown to the public at large. Such a situation prevailed until the mid-1960s. From that time, the cumulative effects of a number of high-level national and international conferences, culminating in the *Study of Critical Environmental Problems* (SCEP) in 1970, produced a growing awareness of global environmental issues. In the SCEP the impact of human activities on regional and global climates was considered, and when it became clear that the issue was of sufficient magnitude to warrant further investigation, a follow-up conference was convened. It focused on inadvertent climate modification, and in 1971 produced a report entitled *The Study of Man's Impact on Climate* (SMIC). The report was recognized as an authoritative assessment of all aspects of human-induced climatic change, and it might even be considered as the final contribution to the 'critical mass' necessary to initiate the numerous and increasingly detailed studies which characterized the next decade and a half.

One of the results of all that activity was the positive identification of human interference as a common element in many of the major problems of the atmospheric environment. The information gathered during the studies is quite variable in content and approach. It has been presented in highly technical scientific reports, as well as in simple, basic articles prepared for popular consumption. The latter are particularly important as a means of disseminating information to the wider audience which past experience suggests must be educated before progress can be made in dealing with environmental problems. Because of the time and space constraints and the marketing requirements of modern journalism, however, the issues are often treated with much less intellectual rigour than they deserve. In addition, the topics are often represented as being new or modern when, in fact, most have existed in the past. Drought, acid precipitation, and the greenhouse effect all result from natural processes, and were part of the earth/atmosphere system even before the human species came on the scene. Their current status, however, is largely the result of human intervention, particularly over the last 200 years, and it is the growing appreciation of the impact of this intervention that has given the issues their present high profile. One topic that can be classified as new is nuclear winter. Unlike the others, which exist at present, and have developed gradually as the accumulated results of a variety of relatively minor inputs, nuclear winter remains in the future, with a potential that can only become reality following the major catastrophic inputs of nuclear war. For many, this is seen as the ultimate intervention – the ultimate blow to environmental equilibrium.

Apart from nuclear winter, present concerns may be seen to some extent as the most recent elements in a continuum. In the 1960s and early 1970s, the main environmental issues were those associated with pollution in its various forms. Consideration was given to the environment as a whole in the academic and scientific community (e.g. Detwyler 1971), but the most pressing popular concerns were often local in origin, dealing with such problems as urban air pollution or reduced water quality in rivers and lakes. Along with increased concern there was also increased understanding of the environment, brought about by the development of educational programmes at all levels – from elementary school to university – and by judicious use of the media by environmental groups such as the Sierra Club, Friends of the Earth, Pollution Probe, and Greenpeace. The high level at which public interest in environmental affairs was sustained during the years when improvement was marginal is in no small measure attributable to these groups.

Public pressure forced the political and industrial establishment to reassess its position on environmental quality. Oil companies, the forest products industry, and even the automobile producers began to express concern for pollution abatement and the conservation of resources. Similar topics began to appear on political platforms, particularly in North America, and although this increased interest was regarded with suspicion – and viewed as a public relations exercise in some quarters – legislation was gradually introduced to alleviate some of the problems. By the early 1970s some degree of control seemed to be emerging. While this may have helped to reduce anxiety over environmental concerns, progress was slow, and some observers attribute the decline in interest in all things environmental at about this time to disenchantment rather than recognition that the problems were being solved (Bach 1972).

Whatever the reason, the level of concern had peaked by then, and when the oil crisis struck in 1973, energy quickly replaced environmental issues in the minds of the politicians, academics, and the public at large. A decade or so later, the energy situation is less critical, the dire predictions of the economists and energy futurists have not come to pass, and the waning of interest in energy topics has been matched by a resurgence of environmental concern, particularly for problems involving the atmosphere. The new issues are global in scale and, at first sight, may appear different from those of earlier years; in fact, they share the same roots. Current topics such as acid rain and the greenhouse effect are linked to the sulphurous urban smogs of two or three decades ago by society's seemingly insatiable demand for more energy, met predominantly by fossil fuels. Population pressures on the land contribute to famine and desertification much as they did in the past. The depletion of the ozone layer, associated with modern chemical and industrial technology, might be considered as only the most recent result of mankind's continual, and seemingly inherent, desire to improve its lot – all the while acting in ignorance of the environmental consequences.

Many of the problems currently of concern have causes which can be traced to ignorance of the workings of the atmosphere. This is particularly true when the impact of air pollution on climate is considered. Almost all human activities produce waste products, and some of these are introduced into the atmosphere. This presented no great problem when populations were small, and technological levels were low, for the atmosphere includes mechanisms designed to keep such emissions in check. For every process adding material to the atmosphere, there is another which works to remove or reduce the excess, either by neutralizing it or by returning it to the earth's surface. Gases, for example, may be absorbed by vegetation, neutralized by oxidation, or dissolved in water and precipitated. Particulate matter falls out of the atmosphere as a result of gravity or is washed out by precipitation. These processes have removed extraneous gases and aerosols from the atmosphere, usually quite effectively, for millions of years.

Ongoing physical and biological activities – such as volcanic eruptions, soil erosion, and the combustion or decay of vegetable matter – ensure that the cleansing process is never complete, and that, in itself, may be part of the system. There are indications, for

9

example, that a minimum level of extraneous material is essential for the working of such atmospheric processes as condensation and precipitation. Thus, a completely clean atmosphere may not be desirable (Barry and Chorley 1987). Desirable or not, it is unlikely to be achieved, given the present rates of gaseous and particulate emissions.

In the past, the main pollutants were natural in origin, and sources such as the oceans, volcanoes, plants, and decaying organic material continue to provide about 90 per cent of the total global aerosol content (Bach 1979). Events such as the eruption of Mount St Helens indicate the continued capability of nature to provide massive volumes of pollutants (Burroughs 1981), but anthropogenic sources are now paramount in many areas. Human activities provide pollutants in such amounts, and with such continuity, that the atmospheric cleansing processes have been all but overwhelmed, and a full recovery may not be possible, even after large-scale attempts to reduce emission levels.

Air pollution was one of the elements which elicited a high level of concern during the heyday of the environmental movement in the late 1960s and early 1970s. It was mainly an urban problem at that time – most common in large cities which had high seasonal heating requirements, were heavily industrialized, had large volumes of vehicular traffic, or experienced combinations of all three. Even then, however, existing air-pollution-control ordinances were beginning to have an effect on the problem. In Pittsburgh, the introduction of smokeless fuel, and the establishment of emission controls on the iron and steel industry, brought a steady reduction in air pollution between 1945 and 1965 (Thackrey 1971). Similar improvements were achieved in London, England, where sunshine levels in the city centre increased significantly following The Clean Air Act of 1956 (Jenkins 1969). The replacement of coal by natural gas as the main heating fuel on the Canadian Prairies, in the 1950s and the 1960s, allowed urban sunshine totals to increase there also (Catchpole and Milton 1976). Success was achieved mainly by reducing the atmospheric aerosol content. Little was done to reduce the gaseous component of pollution, except in California, where, in 1952, gaseous emissions from the state's millions of cars were scientifically proven to be the main source of photochemical smog (Leighton 1966). Prevention of pollution was far from complete, but the obvious improvements in visibility and sunshine totals, coupled with the publicity which accompanied the introduction of new air quality and emission controls in the 1970s, caused the level of concern over urban air pollution to decline markedly by the end of the decade.

10

The relationship between pollution and weather or climate is a complex one. Sometimes climatic conditions will influence the nature and extent of a pollution episode, while at other times the linkages are reversed, allowing the pollutants to instigate or magnify variations in climate. The problem of acid rain illustrates quite well the impact of atmospheric processes on the operation and distribution of a particular group of pollutants, whereas the issues of increased atmospheric turbidity and the depletion of the ozone layer illustrate the other relationship, in which pollutants cause sufficient change in the atmosphere to initiate climatic change.

The full complexity of the earth/atmosphere system is only now beginning to be appreciated, but, in recent years, knowledge of the impact of human social and technological development on the atmospheric environment has grown quite dramatically. Government sponsored studies on such topics as the greenhouse effect and acid rain, along with reports by respected scientists and academics on nuclear winter and the depletion of the ozone layer, have become available. Although often quite technical, they have contained material of sufficient general interest that it could be abstracted and disseminated widely by the media. Other issues have been developed at the popular level from the outset. For instance, the interest in African drought was to a large extent an emotional response to television coverage of events in Ethiopia and the success of the Live Aid concerts of 1985, although excellent academic studies of the problem have been carried out (Bryson and Murray 1977; Glantz 1977).

As more information becomes available on these global environmental issues, it is clear that the present problems have existed undetected for some time. It is also clear that the activities which produced them were entered into with the best of intentions – to improve the quality of human life – and society might even be seen as suffering from its own success. That, of course, does nothing to reduce the seriousness of the problems, but it indicates the need for extreme caution when initiating schemes which promise major advantages. The linkages in the earth/atmosphere system are such that even local or regional changes can be amplified until their impact is felt system-wide. Thus, in the modern world, schemes which might conceivably alter the environment, whether immediately or ultimately, cannot be entered into lightly. Unfortunately, even in the present era of high technology, predicting the eventual reaction of the environment to a specific input is seldom possible, and changes already initiated may well be expanding and intensifying undetected to provide the makings of some future problem.

11

The global environmental topics to be considered in the following chapters are those which currently enjoy a high profile. They include the changing greenhouse effect and the potential impact of nuclear winter, together with various aspects of macro-scale air pollution such as acid precipitation, atmospheric turbidity, and the depletion of the ozone layer. In contrast to these modern, high-tech problems, there is drought – a problem which has plagued mankind for centuries, causing millions of deaths and large-scale environmental degradation, yet remains essentially unsolved. It deserves consideration in its own right but, as a well-established, recurring problem, it also provides a useful contrast with those of more recent origin.

Since society experiences the impact of these elements or induces change in them by way of the atmosphere, an understanding of the workings of that medium is important also. Questions of the composition of the atmosphere, its general circulation, and its role in the global energy budget appear with some regularity in any discussion of these environmental concerns. Consideration of these elements of the atmosphere is therefore a necessary introduction to the issues.

Suggestions for further study

1. Examine the articles dealing with environmental topics in such scientific periodicals as *Science, Scientific American, Nature,* and *New Scientist.* Consider the papers published in 1965, 1975, and 1985 (or in some similar sequence of years) and identify the topics which were most popular during each time period. How do they differ? Is there any indication of present global environmental issues in the earlier time periods?
2. List the laws, statutes, and other ordinances which have been passed in an attempt to improve or maintain environmental quality in your community. How successful have they been? Are existing measures capable of dealing with modern problems in the atmospheric environment?
3. Debate the resolution that:
   Modern technology is at the root of existing global environmental problems, but only technology can provide solutions.

Chapter two

The atmosphere

The atmosphere is a thick blanket of gases, containing suspended liquid and solid particles, which completely envelops the earth, and together with the earth forms an integrated environmental system. As part of this system, it performs several functions which have allowed mankind to survive and develop almost anywhere on the earth's surface. First, it provides and maintains the supply of oxygen required for life itself. Second, it controls the earth's energy budget through such elements as the ozone layer and the greenhouse effect, and – by means of its internal circulation – distributes heat and moisture across the earth's surface. Third, it has the capacity to dispose of waste material or pollutants generated by natural or human activity. Society has interfered with all of these elements, and, through ignorance of the mechanisms involved or lack of concern for the consequences of its action, has created or intensified problems which are now causing concern on a global scale.

The atmospheric gases

The constituents of the atmosphere are collectively referred to as air, although air itself is not a specific gasous element, rather it is a mixture of individual gases each of which retains its own particular properties. Although traces of atmospheric gases have been detected well out into space, 99 per cent of the mass of the atmosphere lies within 30 km of the earth's surface, and 50 per cent is concentrated in the lowest 5 km. Most of the world's weather develops in these lower layers, but certain elements in the upper reaches of the atmosphere are also involved, and some appear as important components in the global environmental issues to be examined here.

Oxygen and nitrogen

Ignoring for the moment the liquids and solids always present, the gaseous mixture which makes up the atmosphere has a remarkably uniform composition throughout. Two gases, oxygen and nitrogen, account for 99 per cent of the total by volume (see Table 2.1). Oxygen (21 per cent by volume) participates readily in chemical reactions, and is one of the necessities of life. It is also capable of absorbing solar radiation. In contrast, nitrogen (78 per cent by volume) is basically inert, seldom becoming directly involved in atmospheric, chemical or biological processes except under extraordinary circumstances. During thunderstorms, for example, the enormous energy flow in a lightning stroke may cause nitrogen to combine with oxygen to produce oxides of nitrogen. On a less spectacular, but ultimately more important level, certain soil bacteria – such as *Clostridium* and *Azobacter* – along with those found in the root nodules of leguminous plants, are capable of fixing the atmospheric nitrogen essential for the creation of the complex nitrogen compounds found in all forms of life on earth (Steila 1976).

Efficient recycling processes maintain the volume of both gases, and turbulent mixing ensures that they are evenly distributed. There is no evidence that the relative levels of oxygen and nitrogen are changing significantly, although there have been measurable changes in the proportions of other gases in the atmosphere. Changes in the nature of oxygen, however, are involved in the depletion of the ozone layer, now recognized as one of the world's major environmental issues. Oxygen can exist in the atmosphere as atomic, diatomic, or triatomic oxygen, depending upon the number of atoms in a molecule. The most common form is

Table 2.1 Average gaseous composition of dry air in the troposphere

Gas	Per cent by volume	Parts per million (ppm)
Nitrogen	78.08	780,840.00
Oxygen	20.95	209,500.00
Argon	0.93	9,300.00
Carbon dioxide	0.0345	345.00
Neon	0.0018	18.00
Helium	0.00052	5.20
Methane	0.00014	1.40
Kryton	0.00010	1.00
Hydrogen	0.00005	0.50
Xenon	0.000009	0.09
Ozone	Variable	Variable

diatomic oxygen (O_2), but the triatomic form called ozone (O_3), created through the combination of the other two types, is present in the upper atmosphere. Ozone is also found close to the surface, on occasion – usually as a product of air pollution – but its main location is in the upper atmosphere, where it effectively filters out short-wave solar radiation at the ultraviolet end of the spectrum. Any change in ozone levels, allowing an increase or decrease in the transmission of radiation, would therefore cause disruption of the earth's energy budget, and lead to alterations in temperature levels and distribution patterns. It is also estimated that a reduction in ozone in the upper atmosphere would allow an increase in the incidence of skin cancer in humans, as well as genetic mutation in lower-level organisms as a result of the increase in the proportion of ultraviolet radiation reaching the earth's surface (Dotto and Schiff 1978).

The release of the fluorocarbon propellant from aerosol spray cans was identified, in the 1970s, as a major cause of the decay of the ozone layer. Emissions from the engines of high-altitude super-sonic transports (SSTs) were also considered potentially damaging. The use of aerosol sprays has declined, the projected high number of SSTs never materialized, and it seemed that threats to the ozone layer had been controlled. More recently, however, evidence of the thinning of the ozone above the Antarctic and the Arctic has renewed concern for its fate.

The minor gases and the greenhouse effect

Although gases other than oxygen or nitrogen account for only about 1 per cent of the atmospheric total, they have an influence quite out of proportion to their volume. The most abundant of these is argon at 0.93 per cent by volume, but it is inert. Another of these minor gases, carbon dioxide, has a much more active role in environmental processes. It comprises only 0.03 per cent by volume, yet is largely responsible for the heating of the atmos-phere, and is a major participant in the process of photosynthesis by which sugars, starches, and other complex organic compounds are produced in plants.

The atmosphere is quite selective in its response to solar radi-ation. It is transparent to high-energy, short-wave radiation, such as that from the sun, but partially opaque to the lower-energy, long-wave radiation emanating from the earth's surface. For example, a major proportion of the radiation in the visible range of the spectrum, between 0.3 and 0.7 micrometers (μm), is trans-mitted through to the surface without losing its high energy content (see Figure 2.1). Once it arrives it is absorbed, the surface heats up, and begins to emit terrestrial long-wave radiation back into the atmosphere. This radiation, from the infrared end of the spectrum – with wavelengths between one and 30 μm – is captured, and the temperature of the atmosphere rises. The capture of the outgoing terrestrial radiation is effected largely by carbon dioxide, along with traces of about another twenty gases, which together are called the greenhouse gases. Water vapour also makes an important contribution to energy absorption in the lower layers of the atmosphere, where it is most densely concentrated. The whole process was labelled the greenhouse effect since the gases, by trapping the heat, appeared to work in much the same way as the glass in a greenhouse. The name remains in common use, although it is now generally accepted that the processes involved are not exactly the same. For example, the glass in the greenhouse acts as a physical barrier to the transfer of energy. There is no such barrier

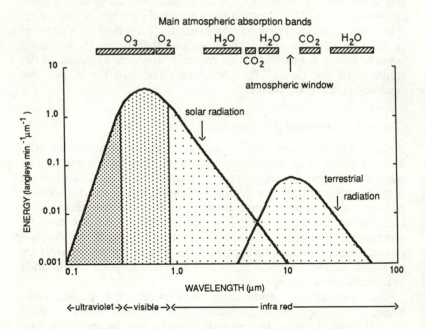

Figure 2.1 Spectral distribution of solar and terrestrial radiation. (Solar radiation is represented by a curve for a black body at 6,000 °K and terrestrial by a black body at 300 °K. A black body is a perfect radiator or absorber of energy.)

in the atmosphere. Whatever the accuracy of the analogy, the selective nature of the atmosphere in its response to radiation is of supreme importance to the earth's energy budget.

Since the greenhouse effect depends upon carbon dioxide and the other gases in the atmosphere, it follows that any change in these gases, including their relative concentration, will have an effect on the intensity of the greenhouse effect. Changes in the proportion of carbon dioxide in the past were brought about by natural processes, but, since the middle of the nineteenth century, human activities have had a major role in increasing the intensity of the greenhouse effect through the production of higher volumes of carbon dioxide. The increasing consumption of fossil fuels contributes directly to the rise in carbon dioxide levels while, indirectly, mankind's destruction of natural vegetation disrupts the carbon cycle, and allows carbon dioxide, which would normally be converted into carbon and atmospheric oxygen, to remain in the atmosphere. Global mean temperatures may already have risen by as much as 2 °C since the beginning of this century, and it is predicted that such a trend will have serious climatological, environmental, economic and perhaps political consequences if allowed to continue, even if only for a few decades. Concerns such as these have promoted the intensification of the greenhouse effect to its present position as a significant environmental issue.

Oxides of sulphur and nitrogen: acid rain

There are many other gases which from time to time become constituents of the atmosphere. These include sulphur dioxide, oxides of nitrogen, hydrogen sulphide, and carbon monoxide, along with a variety of more exotic hydrocarbons, which even in small quantities can be harmful to the environment. These have given rise to local problems in the past, following volcanic eruptions, for example, or associated with urban pollution from industrial or vehicular sources. In recent years, concern has centred on the widespread dissemination and detrimental environmental impact of some of these gases. Increasing industrial activity, and the continued reliance on fossil fuels as energy sources, has caused a gradual, but steady, growth in the proportion of sulphur and nitrogen oxides in the atmosphere over the past two to three decades. These gases, in combination with atmospheric water, are the main ingredients of acid rain, considered by many to be the major environmental problem facing the northern hemisphere at the present time. Acid rain, or acid precipitation as it is more accurately termed, has already damaged aquatic and terrestrial ecosystems in North America and Europe, perhaps irreparably, yet response to the issue has for the most part been half-hearted. Economic and political considerations have tended to override environmental concerns. The southern hemisphere has so far been spared the depredations of acid rain, but as a product of industrial development and a constituent of the atmosphere, it has the potential to spread, and these unaffected areas may not always be immune.

Water in the atmosphere

The creation of acid rain would not be possible without water, another of the major natural constituents of the atmosphere. Lists of the principal gases in the atmosphere – such as Table 2.1 – commonly refer to dry air, but the atmosphere is never completely dry. The proportion of water vapour in the atmosphere in the humid tropics may be as much as 4 per cent by volume, and even above the world's dryest deserts there is water present, if only in fractional amounts. At any one time, the total volume of water in the atmosphere is relatively small, and, if precipitated completely and evenly across the earth's surface, would produce the equivalent of no more than 25 mm of rainfall (Barry and Chorley 1987). In reality, the distribution is very uneven, as a result of regional variability in the dynamic processes which produce precipitation. Intense thunderstorms can yield 25 mm of precipitation in a matter of minutes, whereas the same amount may take several months, or even years, to accumulate under more stable atmospheric conditions. Variations such as these account for annual precipitation totals which range from virtually nothing in some of the world's deserts, to as much as 4000 mm in the monsoon lands of the tropics. Despite regular precipitation in excess of the quantities normally held in the air, the atmosphere never fully dries out. Its moisture is continually replenished by the circulation of water through the earth/atmosphere system by the hydrologic cycle (see Figure 2.2).

Water is unique among the constituents of the atmosphere in that it is capable of existing as solid, liquid, or gas, and of changing readily from one state to another. It becomes involved in energy transfer as a result of these changes. For example, energy absorbed during the conversion of liquid water to water vapour is retained by the latter, in the form of latent heat, until the process is reversed. The stored energy is then released (see Figure 2.3). The water vapour may travel over great distances in the atmosphere in the period between the absorption and re-release of the energy,

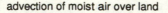

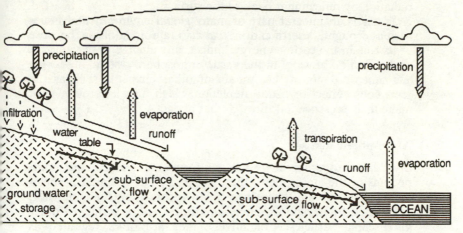

Figure 2.2 The hydrologic cycle

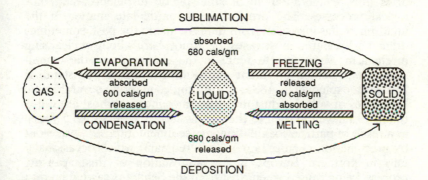

Figure 2.3 Energy transfer during the change in state of water

and in this way energy absorbed in one location is transported elsewhere in the system.

Water is also involved in the earth's energy budget through its ability to absorb and reflect radiation. As vapour, it contributes to the greenhouse effect by absorbing terrestrial radiation, whereas in its liquid and solid forms it can be highly reflective. As clouds in the air or snow on the ground, it may reflect as much as 90 per cent of the solar radiation it intercepts. That radiation, reflected back into space, makes no contribution to the energy requirements of the system. In contrast, clouds can also help to retain terrestrial radiation by reflecting it back to the surface.

Water is an integral part of many global environmental issues such as drought, desertification, and acid rain. In addition, because of its role in the earth's energy budget, any change in the amount or distribution of water in the earth/atmosphere system might well augment or diminish the impact of other elements such as the greenhouse effect or ozone depletion which, in whole or in part, make their presence felt through that budget.

Atmospheric aerosols

In addition to water in its various forms and the gaseous components of the atmosphere, there are also solid or liquid particles dispersed in the air. These are called aerosols, and include dust, soot, salt crystals, spores, bacteria, viruses, and a variety of other microscopic particles. Collectively, they are often regarded as equivalent to air pollution, although many of the materials involved are produced naturally by volcanic activity, forest and grass fires, evaporation, local atmospheric turbulence, and biological processes. The proportion of particulate matter in the atmosphere has increased from time to time in the past, sometimes dramatically, but in most cases the atmosphere's built-in cleansing mechanisms were able to react to the changes, and the overall impact was limited in extent and duration. When the island of Krakatoa exploded in 1883, for example, it threw several cubic kilometres of volcanic dust into the atmosphere. Almost all of it is thought to have returned to the earth's surface in less than 5 years, as a result of particle coagulation, dry sedimentation, and wash-out by precipitation (Ponte 1976). The 'red rain' which occasionally falls in northern Europe is a manifestation of this cleansing process, being caused when dust from the Sahara is carried up into the atmosphere by turbulence over the desert, and washed out by precipitation in more northerly latitudes (Tullett 1984). Thus, the atmosphere can normally cope with the introduction of aerosols by natural processes. Cleansing is never complete, however. There is always a global background level of atmospheric aerosols which reflects a dynamic balance between the output from natural processes and the efficiency of the cleansing mechanisms. Data collected over the past several decades suggest that the level is rising, as a result of the increasing volume of aerosols of anthropogenic origin, although the evidence is sometimes contradictory (Bach 1979).

Measurements since the 1930s – in locations as far apart as

Mauna Loa in Hawaii, Davos in Switzerland, and the Russian Caucasus – show a sharp rise in the atmosphere's aerosol content, or turbidity as it is called. Results from such stations – located at high altitudes and relatively remote from the world's main industrial areas – are considered representative of global background aerosol levels. Recent observations of increasing cold season atmospheric pollution in high latitudes – the so-called 'Arctic haze' – are also considered indicative of rising global levels (Environment Canada 1987). Volcanic activity may also have provided some natural enhancement in recent years, but the close correspondence between elevated turbidity levels and such indicators of human development as industrialization and energy use suggests that anthropogenic sources are major contributors. Some studies claim, however, that the observations are insufficient to allow the human contribution to increased turbidity to be identified (Bach 1979).

Any increase in the turbidity of the atmosphere should cause global temperatures to decline, as the proportion of solar radiation reaching the earth's surface is reduced by scattering and absorption. In addition, the condensation of water vapour around atmospheric aerosols would lead to increased cloudiness and a further reduction in the transmission of incoming radiation. This approach has been used to explain the decline in global average temperatures which occurred between 1940 and 1960, and in the 1970s it was seen by some as the mechanism by which a new ice age would be initiated (Ponte 1976). Such thinking is also central to the concept of nuclear winter which would be caused by a rapid temperature decline following the injection of large volumes of aerosols into the atmosphere (Bach 1986).

Providing a dissenting opinion are those who claim that an increase in atmospheric aerosols would have less serious results. The reduction in insolation is accepted, but it is also considered that there would be a concomitant reduction in the amount of terrestrial radiation escaping into space, which would offset the cooling, and perhaps result in some warming of the lower atmosphere. The overall effects would depend very much on the altitude and distribution of the aerosols (Mitchell 1975).

Contradictory conclusions such as these – which are drawn from the same basic information – are to be expected in climatological studies. They reflect the inadequacy of existing knowledge of the workings of the earth/atmosphere system, and, although research and technological development is changing that situation, it remains a major element in restricting society's response to many global issues.

The vertical structure of the atmosphere

Although its gaseous constituents are quite evenly mixed, the atmosphere is not physically uniform throughout. Variations in such elements as temperature and air pressure provide form and structure in what would otherwise be an amorphous medium. The commonly accepted delineation of the atmosphere into a series of layers, for example, is temperature based (see Figure 2.4). The lowest layer is the troposphere. It ranges in thickness from about 8 km at the poles to 16 km at the equator, mainly as a result of the difference in energy budgets at the two locations, and temperatures within it characteristically decrease with altitude at a rate of 6.5 °C per kilometre. Temperatures at the upper edge of the troposphere average between −50 °C and −60 °C, but in equatorial regions, where it reaches its greatest altitude, values may be as low as −80 °C. The tropospheric lapse rate is, in fact, quite variable, particularly close to the surface. Such variations regularly produce instability in the system, and help to make the troposphere the most turbulent of the atmospheric layers.

The troposphere contains as much as 75 per cent of the gaseous mass of the atmosphere, plus almost all of the water vapour and other aerosols (Barry and Chorley 1987). It is also the zone in which most weather systems develop, and run their course. These

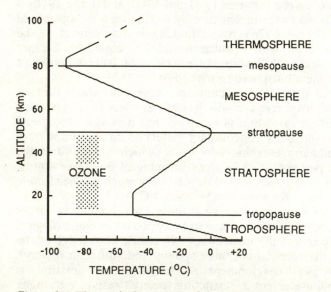

Figure 2.4 The vertical structure of the atmosphere

factors, together with the high level of human intervention in this part of the atmosphere, ensure that many of the global environmental problems of current concern – including acid rain, atmospheric turbidity, and the greenhouse effect – have their origins or reach their fullest extent in the troposphere.

The tropopause marks the upper limit of the troposphere. Beyond it, in the stratosphere, isothermal conditions prevail; temperatures remain constant, at or about the value reached at the tropopause, up to an altitude of about 20 km. Above that level, the temperature begins to rise again – reaching a maximum some 50 km above the surface, at the stratopause, where temperatures close to or even slightly above 0 °C are common. This is caused by the presence of ozone, which absorbs ultraviolet radiation from the sun, and warms the middle and upper levels of the stratosphere, creating a temperature inversion. The combination of that inversion with the isothermal layer in the lower stratosphere creates very stable conditions, and the stratosphere has none of the turbulence associated with the troposphere. This has important implications for the environment. Any foreign material introduced into the stratosphere tends to persist there much longer than it would if it had remained below the tropopause. Environmental problems such as ozone depletion and atmospheric turbidity are aggravated by this situation. Much of the impact of a nuclear winter would result from the penetration of the tropopause by the original explosions, and the consequent introduction of large volumes of pollutants into the stratosphere.

Temperatures again decrease with height above the stratopause and into the mesosphere, falling to as low as −100 °C at the mesopause, some 80 km above the surface. The thermosphere stretches above this altitude, with no obvious outer limit. In this layer, temperatures may exceed 1,000 °C, but such values are not directly comparable to temperatures in the stratosphere and troposphere, because of the rarified nature of the atmosphere at very high altitudes. Knowledge of the nature of the thermosphere, and its internal processes, is far from complete, and linkages between the upper and lower layers of the atmosphere remain speculative. For the climatologist and environmentalist the most important structural elements of the atmosphere are the troposphere and stratosphere. The main conversion and transfer of energy in the earth/atmosphere system takes place within these two layers of the lower atmosphere, and it is through interference in the processes or mechanisms involved that several of the modern problems in the atmospheric environment have become apparent.

The earth's energy budget

Virtually all of the earth's energy is received from the sun in the form of short-wave solar radiation, and balancing this inflow is an equivalent amount of energy returned to space as long-wave terrestrial radiation. The concept is a useful one, but it applies only to the earth as a whole, over an extended time-scale of several years; it is not applicable to any specific area over a short period of time. This balance between incoming and outgoing radiation is referred to as the earth's energy budget.

The earth intercepts only a small proportion of the total energy given out by the sun – perhaps as little as one five-billionth (Critchfield 1983) – and not all of that reaches the earth's surface. The estimates differ in detail, but it is generally accepted that only about 50 per cent of the solar radiation arriving at the outer edge of the atmosphere is absorbed at the surface (Lutgens and Tarbuck 1982). That proportion is split almost equally between direct solar radiation and diffuse radiation, which has been scattered by water vapour, dust, and other aerosols in its passage through the atmosphere (see Figure 2.5). Of the other 50 per cent, some 30 per cent returns to space as a result of reflection from the land and sea, reflection from clouds, or scattering by atmospheric aerosols. The remaining 20 per cent of the original incoming radiation is absorbed by oxygen, ozone, and water vapour in the atmosphere. Thus, of every 100 units of solar energy arriving at the outer edge of the atmosphere, 70 are absorbed into the earth/atmosphere system and 30 are returned to space having made no contribution to the system. Most of the 50 units absorbed by the earth are re-radiated as long-wave terrestrial radiation. Some energy is also transferred into the atmosphere by convective and evaporative processes at the earth's surface. The greenhouse gases trap the bulk of the outgoing energy, but the atmosphere is transparent to wavelengths between 8 μm and 11 μm, and this allows 5 units of radiation to escape directly to space through the so-called atmospheric window. Some of the terrestrial radiation absorbed by the atmosphere is emitted to space also, but sufficient accumulates to allow 95 units to be re-radiated back to the surface. The exchange of energy between the atmosphere and the earth's surface involves amounts apparently in excess of that provided by solar radiation. This is a direct function of the greenhouse effect, and its ability to retain energy in the lower atmosphere. Eventually all of the long-wave radiation passes out into space also, but not before it has provided the energy necessary to allow the various atmospheric and terrestrial processes to function.

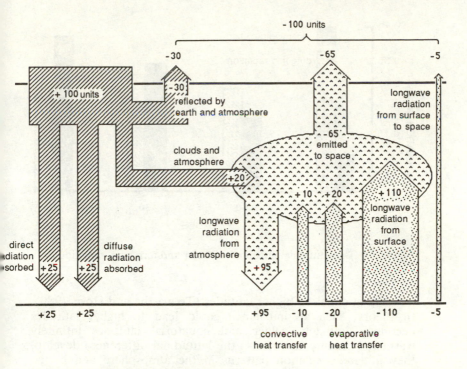

Figure 2.5 The earth's energy budget

The effects of some of the earth's environmental problems are experienced by way of the energy budget. The depletion of the ozone layer allows additional radiation to reach the surface while increases in the level of carbon dioxide in the atmosphere encourage the retention of terrestrial radiation, and changes in turbidity disrupt both incoming and outgoing radiation. Nuclear war poses the greatest threat to the earth's energy budget; it would destroy present patterns of energy flow into and out of the atmosphere, and create nuclear winter.

The concept of balance in the earth's heat budget is a useful one, but it provides only a global picture, and cannot be applied to specific areas. There is a definite latitudinal imbalance in the budget. Annually, the equator receives about five times the amount of solar radiation reaching the poles, and those areas equatorwards of 35 degrees of latitude receive more energy than is returned to space (see Figure 2.6). The excess of outgoing radiation over incoming – poleward of 35 degrees of latitude – creates

25

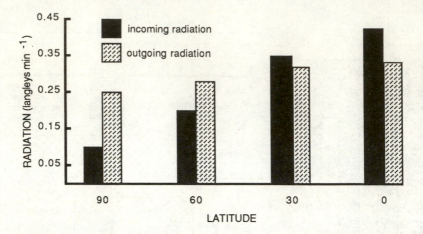

Figure 2.6 The latitudinal imbalance in solar and terrestrial radiation

a radiation deficit in higher latitudes (Trewartha and Horn 1980). In theory, such an imbalance could lead to higher latitudes becoming infinitely colder and equatorial latitudes infinitely warmer. In reality, as soon as the latitudinal differences develop, they initiate circulation patterns in the atmosphere and in the oceans, which combine to transfer heat from the tropics towards the poles, and in so doing serve to counteract the imbalance.

Oceanic and atmospheric circulation patterns

The circulation of the oceans

Both systems are quite intimately linked. The prevailing winds in the atmospheric circulation, for example, drive water across the ocean surface at speeds of less than 5 km per hour, in the form of broad, relatively shallow drifts. In some cases they carry warm water polewards, in others they carry cooler water into lower latitudes (see Figure 2.7). In addition, density differences, in part thermally induced, cause horizontal and vertical movement of water within the oceans. All of these processes help to transfer excess radiation from equatorial regions towards the poles. This is illustrated particularly well in the North Atlantic, where the warm waters of the Gulf Stream Drift ensure that areas as far north as the Arctic Circle are anomalously mild during the winter months. Estimates of poleward energy transfer in the northern hemisphere

26

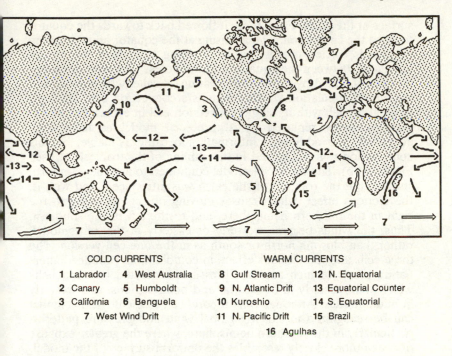

COLD CURRENTS		WARM CURRENTS	
1 Labrador	4 West Australia	8 Gulf Stream	12 N. Equatorial
2 Canary	5 Humboldt	9 N. Atlantic Drift	13 Equatorial Counter
3 California	6 Benguela	10 Kuroshio	14 S. Equatorial
	7 West Wind Drift	11 N. Pacific Drift	15 Brazil
			16 Agulhas

Figure 2.7 The circulation of the oceans

indicate that ocean transfer exceeds atmospheric transfer in low latitudes, whereas the latter is more important in mid to high latitudes. On average, oceanic transport accounts for 40 per cent of the total energy transfer, and atmospheric transport for 60 per cent (Trewartha and Horn 1980)

The circulation of the atmosphere

George Hadley developed the classic model of the general circulation of the atmosphere some 250 years ago. It was a simple convective system, based on the concept of a non-rotating earth with a uniform surface, which was warm at the equator and cold at the poles (see Figure 2.8a). The warmth caused the surface equatorial air to become buoyant and rise vertically into the atmosphere. As it rose away from its source of heat, it cooled and became less buoyant, but was unable to sink back to the surface because of the warm air rising behind it. Instead, it spread north and south away from the equator, eventually returning to the surface at the poles. From there it flowed back towards the equator to close the circulation. The air rising at the equator and spreading polewards carried energy with it, helping to reduce the energy imbalance between the equator and the poles. This type of energy transfer, initiated by differential heating, is called convection, and the closed circulation which results is a convection cell. Hadley's original model, with its single convection cell in each hemisphere, was eventually replaced by a three-cell model as technology advanced, and additional information became available, but his contribution was recognized in the naming of the tropical cell (see Figure 2.8b). The three-cell model continued to assume a uniform surface, but the rotation of the earth was introduced, and with it, the Coriolis effect, which causes moving objects to swing to the right in the northern hemisphere, and to the left in the southern. Thus, the winds became westerly or easterly in this new model, rather than blowing north or south as in the one-cell version. The three cells and the Coriolis effect, in combination, produced alternating bands of high and low pressure, separated by wind belts which were easterly in equatorial and polar regions, and westerly in mid latitudes. Although only theoretical, elements of this model can be recognized in existing global wind and pressure patterns, particularly in the southern hemisphere, where the greater expanse of ocean more closely resembles the uniform surface of the model.

In the late 1940s and 1950s, as knowledge of the workings of the atmosphere increased, it became increasingly evident that the three-cell model oversimplified the general circulation. The main problems arose with the mid-latitude cell. According to the model, the upper airflow in mid-latitudes should have been easterly, but observations indicated that it was predominantly westerly. The winds followed a circular path centred on the pole, which led to their description as circumpolar westerlies. The observations also indicated that most energy transfer in mid-latitudes was accomplished by horizontal cells rather than the vertical cell indicated by the model. The mechanisms involved included travelling high- and low-pressure systems at the surface, plus wave patterns in the upper westerlies called Rossby waves (Starr 1956). Modern interpretations of the general circulation of the atmosphere retain the tropical Hadley cell, but the horizontal eddies have come to dominate mid-latitudes, and have even replaced the simple thermal cell of polar latitudes (Barry and Chorley 1987).

The flow pattern adopted by the Rossby waves is quite variable, but that variability makes a major contribution to energy transfer. When the westerlies follow a latitudinal path from west to east they are said to be zonal, and the strength of the latitudinal flow is

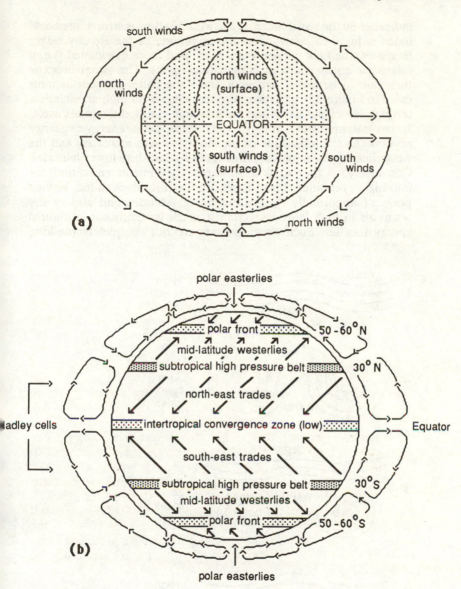

Figure 2.8 (a) Simple convective circulation on a uniform, non-rotating earth, heated at the equator and cooled at the poles; (b) Three-cell model of the atmospheric circulation on a uniform, rotating earth heated at the equator and cooled at the poles

29

indicated by the zonal index. If the westerly flow is strong, the zonal index is high. In contrast, as the amplitude of the Rossby waves increases, the flow becomes less zonal and more meridianal (i.e. it follows a north–south or longitudinal path). The zonal index is then said to be low. Changes in the wave patterns occur as indicated in Figure 2.9, and the entire sequence from high-zonal index, through low-zonal index and back to high is called the index cycle. It has an important role in the atmospheric energy-exchange process. As the amplitude of the Rossby waves increases, and the westerlies loop southwards, they carry cold air into lower latitudes. Conversely, as they loop northwards, they introduce warmer air into higher latitudes. These loops are often short-circuited, leaving pools of abnormally cool air in lower latitudes and abnormally warm air in high latitudes. The net result is significant latitudinal energy transfer. Such developments are not completely random,

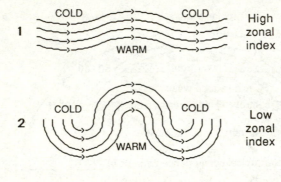

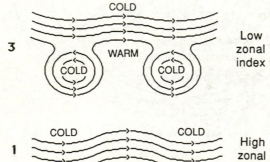

Figure 2.9 The index cycle associated with the meandering of the mid-latitude westerlies in the northern hemisphere

but neither are they predictable. The cycle occurs over a period of 3 to 8 weeks, and is repeated with some frequency, yet it lacks the regularity necessary for forecasting.

The jet streams

The modification of the three-cell model in the 1940s and 1950s was made possible in large part by improved knowledge of conditions in the upper atmosphere. The upper atmospheric circulation is quite complex in detail, but in general terms it is characterized by an easterly flow in the tropics and a westerly flow in mid to high latitudes. Within these broad airflows, there are relatively narrow bands of rapidly moving air called jet streams, in which wind speeds average 125–130 km per hour in places. The jet streams are usually located at the tropopause, and are associated with zones in which steep temperature gradients exist, and where, in consequence, the pressure gradient is also steep. The most persistent jets are found in the sub-tropics, at the poleward edge of the tropical Hadley cell, and in mid-latitudes at the polar front (see Figure 2.10). Both of these jets generally flow from west to east. In addition, an Arctic jet, associated with the long polar night, has been identified (Hare and Thomas 1979), and an intermittent, but recognizable, easterly jet is a feature of the upper atmospheric circulation in equatorial regions (Barry and Chorley 1987).

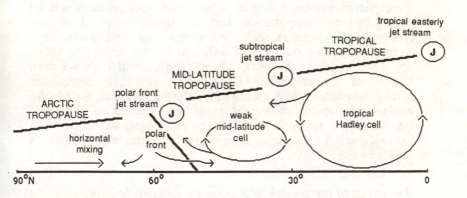

Figure 2.10 Schematic diagram of the vertical circulation of the atmosphere and the location of the major jet streams in the northern hemisphere

The northern hemisphere polar-front jet stream, the one most commonly encountered as head winds or tail winds during trans-continental or trans-oceanic flights, is the best known of all the jet streams. It circles the earth in mid-latitudes, following a meandering track from west to east at speeds averaging perhaps 100 km per hour, but with a maximum recorded speed of almost 500 km per hour (Eagleman 1985). During the winter it follows a more southerly track, close to 35° N, and has an average velocity of 130 km per hour, whereas in the summer it is located closer to 50° N, and its velocity decreases to about 65 km per hour.

The influence of the polar-front jet stream extends to the lower atmosphere, through its control over the various systems which produce the surface weather conditions. For example, the difference between a mild winter and a cold one, in the interior of North America, is often determined by the location of the polar-front jet stream. A more southerly track allows the cold polar air on the north side of the jet to penetrate into lower latitudes, whereas a more northerly track allows the continent to remain bathed in the warmer, southern air. The jet also exerts its influence on moisture regimes – through its control over the tracks followed by mid-latitude, low-pressure systems – and it has been implicated in the tornado outbreaks which occur in North America every spring. North–south thermal contrasts are strong at that time, and the jet is therefore particularly vigorous (Eagleman 1985).

The importance of the jet stream and the associated upper westerlies, from an environmental point of view, lies in their ability to transport pollutants over great distances through the upper atmosphere. Smoke, volcanic debris, and acid particles are all spread by such transportation, and, as a result, the problems they represent are global in scale. When above-ground atomic tests were being carried out in the USSR and China, during the 1950s and early 1960s, radioactive fall-out was carried over northern Canada in the jet stream (Hare 1973). A similar mechanism spread the products of the Chernobyl nuclear accident. Any future nuclear war would cause great quantities of debris to be thrown into the upper atmosphere, where the jet streams would ensure a hemispheric distribution and contribute to the rapid onset of nuclear winter.

The effects of surface conditions and seasonal variations

Modern representations of the general circulation of the atmosphere take into account the non-uniform nature of the earth's surface, with its mixture of land and water, and include consider-

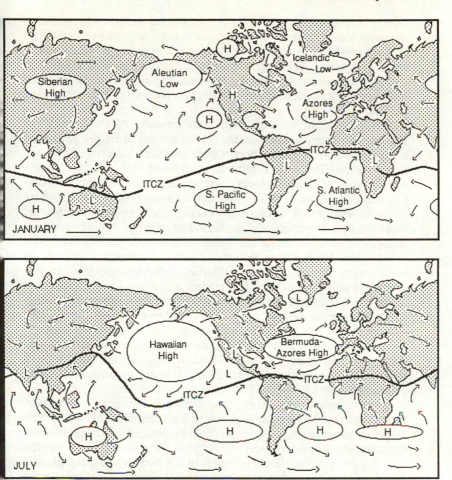

Figure 2.11 Global wind and pressure patterns for January and July

ation of seasonal variations in energy flow (see Figure 2.11).

Land and water respond differently to the same energy inputs, because of differences in their physical properties. Land tends to heat up and cool down more rapidly than water, and temperatures over land exhibit a greater range, diurnally and seasonally, than those over water. These temperature differences in turn have an impact on atmospheric pressure, particularly in the northern hemisphere with its juxtaposition of oceans and major landmasses.

33

During the northern summer, for example, higher temperatures promote lower pressure over the continents, whereas the adjacent seas are cooler, and pressure remains high in the cells which represent the sub-tropical high pressure belt over the North Atlantic and Pacific oceans. By altering the regional airflow, such pressure differences cause disruption of the theoretical, global-circulation patterns.

The permanent or semi-permanent features of the circulation also vary in extent and intensity, by the season or by the year, and in some cases over much shorter time-scales as patterns of energy-flow change. There is a general southward shift of the systems during the northern winter, for example. The Inter-tropical Convergence Zone (ITCZ) – the belt of low pressure produced by the combination of equatorial heating and the convergence of trade winds – migrates southwards in sympathy with the southward movement of the zone of maximum-energy receipt. Behind it, the sub-tropical anticyclones also move south, contracting as they do so, allowing the mid-latitude westerlies and their associated travelling highs and lows to extend their influence into lower latitudes.

In the southern hemisphere at this time, the sub-tropical high pressure systems expand, and extend polewards over the oceans. The increased solar radiation of the summer season contributes to the formation of thermal lows over South America, Africa, and Australia. The absence of major landmasses polewards of 45–50° S allows the westerly wind-belt to stretch as a continuous band around the earth.

The patterns change during the northern summer, as the ITCZ moves northwards again, and the other elements of the system respond to changing, regional, energy budgets. Although such changes are repeated year after year, the movements of the systems are not completely reliable. In the tropics, for example, the ITCZ migrates at different rates and over different distances from one year to the next. This inherent variability contributes to the problem of drought. It also adds complexity to the impact of environmental problems – such as the enhanced greenhouse effect or atmospheric turbidity – which cause changes in the earth's energy budget, and therefore have the potential to alter circulation patterns.

Autovariations and feedback mechanisms

It is convenient to consider the various elements of the earth/atmosphere system as separate entities, as has been done here. They are, however, quite intimately linked, and understanding the nature of these links is important – not only in the pure scientific study of the atmosphere itself, but also in the applied, interdisciplinary approach required for the study of modern global environmental problems. The elements and processes incorporated in the earth's energy budget and the atmospheric circulation, for example, are parts of a dynamic system, in which the components are sufficiently integrated that one change will automatically produce others. Such changes, produced through the activities of internal processes, are called autovariations (Trewartha and Horn 1980), and, in combination with feedback mechanisms, they may augment or dampen the effects of a particular change in the system. Lower temperatures at the earth's surface, for example, would allow the persistence of snow cover beyond the normal season. This, in turn, would increase the amount of solar radiation reflected back into space, causing surface temperatures to fall even more, and would encourage the snow to remain even longer. Such a progression, in which the original impact is magnified, illustrates the concept of positive feedback. Rising temperatures may initiate negative feedback. One of the initial effects of higher temperatures would be an increase in the rate of evaporation from the earth's surface. Subsequent condensation of the water vapour in the atmosphere would increase cloudiness, and would reduce the amount of solar radiation reaching the surface. As a result, temperatures would fall again and the initial impact of rising temperatures would be diminished. Ultimately, these changes in the earth's energy budget would be reflected in the general circulation of the atmosphere. Many current global issues – such as the intensification of the greenhouse effect, increased atmospheric turbidity, and desertification – involve autovariations and include positive or negative feedback in their development.

Summary

Relationships between the components of current global environmental problems and the physical elements of the atmosphere are now well established. The next logical step is to use this information to forecast future developments, so that steps may be taken to minimize environmental impact. This is proving to be no easy task. The ultimate causes and effects of these relationships remain imperfectly understood, despite major advances in the acquisition and analysis of atmospheric data. As a result, long-range forecasting is difficult. This has important implications for the study of these global issues in which climate is involved and restricts the responses to the problems which arise.

Suggestions for further study

1. The eruption of Mount St Helens in 1980 was widely reported in popular scientific journals. Using *National Geographic* (159: 1981), *Scientific American* (244: 1981), *New Scientist* (89: 1980), and similar publications, prepare a table listing the gases given off during the eruption. Compare these with the normal atmospheric gases listed in Table 2.1. How many appear on both lists? On an outline map of the world indicate the spread of volcanic dust following the eruption. Explain, as far as possible, the final distribution of the dust using the winds and pressure belts of the general circulation of the atmosphere. What impact would an increased frequency of volcanic eruptions and the consequent increase in emissions of gas and dust have on world climates?

2. Write a short account of the contribution of George Hadley, William Ferrel, and Carl Rossby to the study of atmospheric circulation patterns.

3. Feedback mechanisms in the atmosphere may reinforce or dampen the impact of a particular change. List examples of positive and negative feedback mechanisms in the earth/atmosphere system. Try to identify those which contribute to current global environmental issues. The following sequence is often used as a good example of positive feedback in the system: lower temperatures → increased snow cover → higher albedo → lower temperatures. How does it work? Could such a series of developments lead to the onset of an Ice Age?

4. For a number of meteorological stations in your area obtain annual data on solar radiation received or, if that is unavailable, the number of sunshine hours recorded. Try to explain the differences among the stations in terms of the factors which influence the receipt of solar radiation at the earth's surface. How important do you think local factors, such as differences between urban and rural or industrial and non-industrial sites, might be?

Chapter three

Drought, famine, and desertification

In recent years the world's attention has been drawn to the Third World nations of Africa by the plight of millions of people who are unable to provide themselves with food, water, and the other necessities of life. The immediacy of television, with its disturbing images of dull-eyed, pot-bellied, malnourished children, skeletons of cattle in dried-up water courses, and desert sands relentlessly encroaching upon once productive land, raised public concern to unexpected heights, culminating in the magnanimous response to the Live Aid concerts of 1985. The requirements of modern popular journalism and broadcasting has meant that coverage of the situation has often been narrow, highly focused, and shallow – lacking the broader, deeper investigation necessary to place the events in a geographical or environmental framework. For example, the present problems are often treated as modern phenomena, although they are indigenous to the areas involved. The inhabitants of sub-Saharan Africa have suffered the effects of drought and famine for thousands of years. It is part of the price that has to be paid for living in a potentially unstable environment. The current episode seems particularly catastrophic, but it is only the most recent in a continuing series. In dealing with the situation, the media have also tended to concentrate on the problems of famine and its consequences, while most of the aid being supplied has, of necessity, been aimed at alleviating hunger. However, the famine, which is the direct cause of the present suffering and hardship, is a symptom of more fundamental problems. Elements of a cultural, socio-economic, and political nature may contribute to the intensity and duration of the famine, but, in areas such as the Sahel, the ultimate causes are to be found in the environment, working through such elements as drought and desertification. The former, through its impact on plants and animals, destroys the food supply and initiates the famine; the latter, with its associated environmental changes, causes productive land to become barren desert, and ensures that the famine will persist, or at least recur with some frequency. Together, drought, famine, and desertification have been implicated in some of the major human catastrophes of the past and present, and will undoubtedly contribute to future suffering and despair.

Drought

The problem of definition

Drought is a rather imprecise term with both popular and technical usage. To some, it indicates a long, dry spell, usually associated with lack of precipitation, when crops shrivel and reservoirs shrink. To others, it is a complex combination of meteorological elements, expressed in some form of moisture index. There is no widely accepted definition of drought. It is, however, very much a human concept, and many current approaches to the study of drought deal with moisture deficiency in terms of its impact on human activities, particularly those involving agriculture. Agricultural drought is defined in terms of crop growth or development. This in turn may lead to economic definitions of drought when, for example, dry conditions reduce yield or cause crop failure, leading to a reduction in income. It is also possible to define drought in purely meteorological terms, where moisture deficiency is measured against normal or average conditions, which have been established through long-term observation (Katz and Glantz 1977).

Aridity and drought

The establishment of normal moisture levels also allows a distinction to be made between aridity and drought. Aridity is usually considered to be the result of low average rainfall, and is a permanent feature of the climatology of a region (see Figure 3.1a). The deserts of the world, for example, are permanently arid, with rainfall amounts of less than 100 mm per year. In contrast, drought is a temporary feature, occurring when precipitation falls below normal or when near normal rainfall is made less effective by other weather conditions such as high temperature, low humidity, and strong winds (Felch 1978).

Aridity is not a prerequisite for drought. Even areas normally considered humid may suffer from time to time, but some of the worst droughts ever experienced have occurred in areas which include some degree of aridity in their climatological makeup.

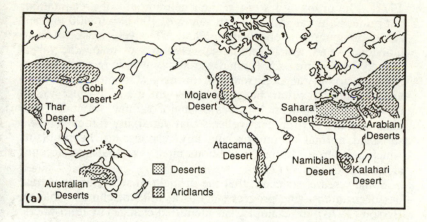

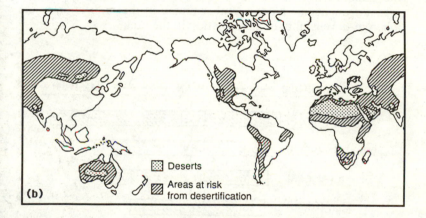

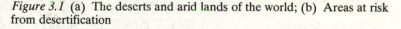

Figure 3.1 (a) The deserts and arid lands of the world; (b) Areas at risk from desertification

Source: Compiled from various sources

Along the desert margins in Africa, for example, annual precipitation is low, ranging between 100 mm and 400 mm. Under normal conditions, this would allow sufficient vegetation growth to support pastoral agriculture, and some arable activity might also be possible, if dry-farming techniques were employed. Drought occurs with considerable regularity in these areas (Le Houérou

39

1977). The problem lies not in the small amount of precipitation, but rather in its variability. Mean values of 100 mm to 400 mm are based on long-term observations, and effectively mask totals in individual years, which may range well above or below the values quoted (see Figure 3.2). Rainfall variability is now recognized as a major factor in the occurrence of drought (Oguntoyinbo 1986), and a number of writers have questioned the use of 'normal' values in such circumstances. In areas of major rainfall variability, the nature of the environment reflects that variability rather than the so-called normal conditions, and any response to the problems which arise from drought conditions must take that into account (Katz and Glantz 1977).

Some researchers claim that the drought in sub-Saharan Africa has been intensifying progressively as a result of climatic change. Bryson (1973), for example, has identified changes in atmospheric circulation patterns which could intensify and prolong drought in the Sahel. There is, however, no conclusive evidence that the recent droughts are anything other than a further indication of the inherent unreliability of precipitation in the area.

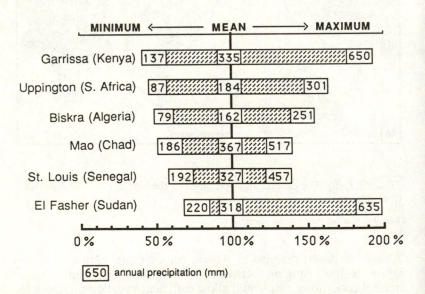

Figure 3.2 Variability of precipitation at selected stations in Africa

Source: Compiled from data in Landsberg (1986)

The human response to drought

Each generation has its images of drought. In the 1980s it is Ethiopia; in the 1960s it was the Sahel; in the 1930s it was the Dustbowl, described with such feeling by John Steinbeck in *The Grapes of Wrath*. Although these have gripped the popular imagination, they are only the more serious examples of perhaps the most ubiquitous climatological problem that society has to face. In some areas of Asia and Africa, drought has been part of the way of life from time immemorial; elsewhere it is highly irregular and, as a result, all the more serious. Such was the case in 1976, when the normally humid UK was sufficiently dry that the government felt it necessary to appoint a Minister of Drought.

From a human perspective, drought may be seen as a technological problem, an economic problem, a political problem, a cultural problem, or sometimes a multi-faceted problem involving all of these. Whatever else it may be, however, it is always an environmental problem, and basic to any understanding of the situation is the relationship between society and environment in drought-prone areas.

Over thousands of years, certain plants and animals have adapted to life with limited moisture. Their needs are met, therefore no drought exists. This is the theoretical situation in most arid areas. In reality, it is much more complex, for although the flora and fauna may exist in a state of equilibrium with other elements in the environment, it is a dynamic equilibrium, and the balance can be disturbed. Changes in weather patterns, for example, might cause dry areas to become even drier. If the plants and animals can no longer cope with the reduced water supply, they will suffer the effects of drought. Depending upon the extent of the change, plants might die from lack of moisture, they might be forced out of the area as a result of competition with species more suited to the new conditions, or they might survive, but at a reduced level of productivity. The situation is more complex for animals, but the response is often easier. In addition to requiring water, they also depend upon the plants for food, and their fate, therefore, will be influenced by that of the plants. They have one major advantage over plants, however. Being capable of movement, they can respond to changing conditions by migrating to areas where their needs can be met. Eventually some degree of balance will again be attained, although certain areas – such as the world's desert margins – can be in a continual state of flux for long periods of time.

The human animal, like other species, is also forced to respond

41

to such changing environmental conditions. In earlier times this often involved migration, which was relatively easy for small primitive communities, living by hunting and gathering, in areas where the overall population was small. As societies changed, however, this response was often no longer possible. In areas of permanent, or even semi-permanent agricultural settlement, with their associated physical and socio-economic structures, migration was certainly not an option – indeed, it was almost a last resort. The establishment of political boundaries, which took no account of environmental patterns, also restricted migration in certain areas. As a result, in those regions susceptible to drought, the tendency (perhaps even the necessity) to challenge the environment grew. If sufficient water was not available from precipitation, either it had to be supplied in other ways – by well and aquaduct, for example – or different farming techniques had to be adopted to reduce the moisture need in the first place. The success of these approaches depended very much on such elements as the nature, intensity, and duration of the drought, as well as on the numbers, stage of cultural development, and technological level of the peoples involved.

Types of drought

C.W. Thornthwaite, the eminent applied climatologist, whose pioneering water-balance studies made a major contribution to the understanding of aridity, recognized four types of drought, defined in terms of agricultural requirements (Thornthwaite 1947). These were permanent, seasonal, contingent, and invisible drought. Agriculture is not normally possible in areas of permanent drought, since there is insufficient moisture for anything but the xerophytic plants which have adapted to the arid environment. Crops can be produced in such areas, but only at great expense or under exceptional circumstances – such as those which apply to the Israeli activities in the Negev Desert, for example (Berkofsky 1986).

On the margins of the world's great deserts, there are regions of seasonal drought – where arid conditions prevail for part of the year, but which are balanced by a distinct wet season. Much of India, the Sahel, and the southern parts of Africa experience such seasonal drought. Agriculture is carried out, often very successfully, during the wet season, and even during the dry season if the moisture from the preceding rainy season can be retained. If for some reason the rainy season is curtailed, however, the potential for drought is great, and it is not surprising that areas such as the Sahel and the Indian sub-continent have experienced some of the world's most spectacular and catastrophic droughts. The problem is intensified in drier years by the irregularity with which the rains fall, making planning difficult if not impossible.

Irregular and variable precipitation is also characteristic of contingent drought. In Thornthwaite's definition, this is experienced in areas which normally have an adequate supply of moisture to meet crop needs. Serious problems arise because the agricultural system is not set up to cope with unpredictable and lengthy periods of inadequate precipitation. The interior plains of North America have suffered from contingent drought for hundreds of years, and the drought of 1975–6 in Britain would fit this category also.

The presence of these three types of drought is indicated by physical changes in the soil and vegetation in the areas affected, but there is also a fourth type which is less obvious. This is the so-called invisible drought, which often can be identified only by sophisticated instrumentation and statistical techniques. The crops appear to be growing well, even to the experienced observer, and there is no obvious lack of precipitation. However, moisture requirements are not being met, the crop is not growing at its optimum rate, and the potential yield from the land is reduced. Invisible drought can be dealt with relatively easily by irrigation. In eastern Britain, for example, supplementary moisture has been supplied to sugar beet and potato crops since at least the late 1950s to deal with that problem (Balchin 1964).

Thornthwaite's treatment of the measurement and classification of drought is only one of many. It may not meet all needs, but its broadly climatological approach lends itself to the geographical examination of drought-prone environments. The greatest human impact is felt in those areas which experience seasonal or contingent drought, although the nature of the impact is different in each case. The influence of the other two types of drought is limited. In areas of permanent drought, for example, populations are small, and may exist only under special circumstances – such as those at an oasis – where the effects of the drought are easily countered. Although invisible drought may have important consequences for individuals, it often passes unrecognized. It does not produce the life and death concerns that prevail in areas of seasonal drought, nor does it have the dire economic impacts that may be experienced by the inhabitants of areas of contingent drought.

Seasonal drought in the sub-tropics

Seasonal drought is most commonly experienced in the sub-tropics. There, the year includes a distinct dry season and a distinct wet season, associated with the north–south movement of the inter-tropical convergence zone (ITCZ) and its attendant wind and pressure belts (see Figure 2.10).

During the dry season, these areas are dominated by air masses originating in the sub-tropical, high pressure systems – which characteristically contain limited moisture, and are dynamically unsuited to produce much precipitation. Anticyclonic subsidence prevents the vertical cloud development necessary to cause rain. In contrast, the rainy season is made possible by the migration of the ITCZ, behind which the combination of strong convection and air mass convergence promotes the instability and strong vertical growth which leads to heavy rainfall. The passage of the ITCZ – in Africa, India, south-east Asia, and Australia – allows the incursion of moist, relatively unstable air from the ocean over the land to initiate the wet season. This is the basis of the monsoon circulation in these areas, and it is often the failure of this circulation that sets up the conditions necessary for drought. If, for example, the ITCZ fails to move as far polewards as it normally does, those regions which depend upon it to provide the bulk of their yearly supply of water will remain under the influence of the dry air masses and receive little or no rainfall. Similarly, any increase in the stability of the airflow following the passage of the ITCZ will also cause a reduction in water supply. It is developments such as these that have set the stage for some of the worst droughts ever experienced.

Drought and famine in Africa

Although seasonal drought is experienced in all of the world's sub-tropical areas, in recent years the greatest effects have been felt in Africa, specifically in the region known as the Sahel. The Sahel proper is that part of western Africa lying to the south of the Sahara Desert and north of the tropical rainforest. It comprises six nations, stretching from Senegal, Mauritania, and Mali in the west, through Burkina Faso, to Niger and Chad in the east. This region, with its population of 33 million inhabiting slightly more than 5 million sq. km of arid or semi-arid land, came to prominence between 1968 and 1973 when it was visited by major drought, starvation, and disease. Despite this prominence, it is in fact only part of a more extensive belt of drought-prone land in Africa south of the Sahara. Drought pays no heed to political boundaries, reaching as it does to the Sudan, Ethiopia, and Somalia in the east and including the northern sections of Ghana, Nigeria, Cameroon, the Central African Republic, Uganda, and Kenya (see Figure 3.3).

The atmospheric circulation in sub-Saharan Africa

The supply of moisture in all of these areas is governed by seasonal fluctuations in the position of the ITCZ. Dry conditions are associated with hot, continental tropical (cT) air, from the Sahara in the west and the Arabian Peninsula in the east, which moves in behind the ITCZ as it migrates southwards during the northern hemisphere's winter. At its most southerly extent, in January or February, the ITCZ remains at about 8 degrees north of the equator in West Africa, but curves sharply southwards across the centre of the continent to reach 15–20° S in East Africa (see Figure 3.4). Apart from East Africa, which receives some precipitation brought in off the Indian Ocean by the north-east trade winds of the winter monsoon, most of the northern part of the continent experiences its dry season at that time. All of West Africa, beyond a narrow strip some 200 km wide along the coast, is under the influence of a north-easterly airflow, from the central Sahara. This is known locally as the Harmattan – a hot, dry wind, which carries with it large volumes of fine dust from the desert.

The ITCZ moves north again during the northern summer, and by July and August has reached its most northerly location at about 20° N (see Figure 3.4). Hot, moist, maritime tropical (mT) air flows in behind it, bringing the rainy season. In the east, the moisture is provided by air masses from the Indian Ocean as part of the Asiatic monsoon system, while in the west it arrives in the south-westerly flow off the South Atlantic. Although relatively simple to describe in general terms, in detail the precipitation patterns are quite complex. In West Africa, for example, the existence of four weather zones, aligned east–west in parallel with the ITCZ, was recognized by Hamilton and Archbold in 1945, and these, with some modern modifications, (e.g. Musk 1983) provide the standard approach to the regional climatology of the area (see Figure 3.5). Each of the zones is characterized by specific weather conditions, which can be identified in the precipitation regimes of the various stations in the area (see 'Suggestions for further studies' on p. 66). The precipitation is caused mainly by convection and convergence, and reaches the ground in a variety of forms, ranging from light, intermittent showers, to the heavy downpours associated with violent thunderstorms or line squalls.

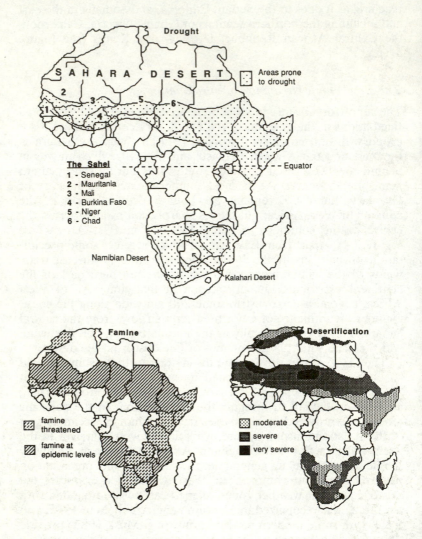

Figure 3.3 The distribution of drought, famine, and desertification in Africa

Source: After Canadian International Development Agency (1985)

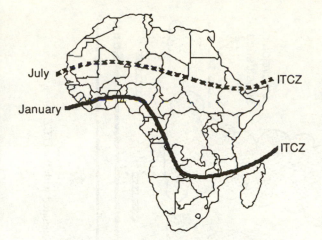

Figure 3.4 Seasonal changes in the position of the ITCZ in Africa

The easterly tropical jet is also active in the upper atmosphere at this time, and may well influence the amount, intensity, and distribution of precipitation (Kamara 1986).

Drought and human activity

Throughout sub-Saharan Africa, the peak of the rainy season in July and August is also the time of year when grass and other forage is most widely available for the herds of cattle, camels, goats, and sheep belonging to the local agriculturalists. The original herdsmen lived a nomadic existence, following the rains north in the summer and south in the winter, to obtain the food their animals needed. In the southern areas there was sufficient moisture available to allow a more permanent life-style, supported by basic agriculture producing sorghum and millet. Drawn south by the rains, the herdsmen eventually encroached upon this farmed land, but instead of the conflict that might have been expected in such a situation, the two societies enjoyed a basic symbiotic relationship. The nomads exchanged meat and milk for grain; the cattle grazed the stubble, and provided natural fertilizer for the following year's crop.

Although the movement of the ITCZ is repeated season after season, it shows some irregularity in its timing, and in the latitudinal distance it covers each year. The rainfall associated with it is never completely reliable, and this is particularly so in those

47

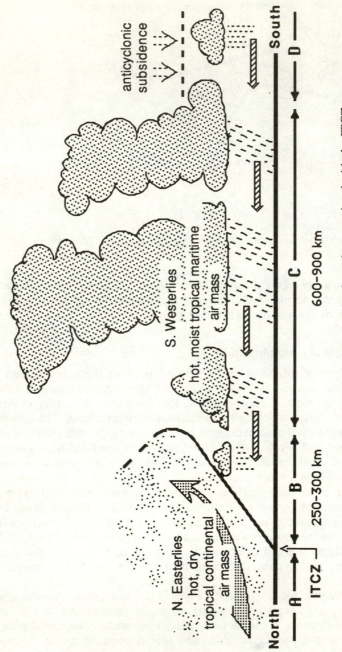

Figure 3.5 North–south section across West Africa, showing bands of weather associated with the ITCZ

Source: After Hamilton and Archbold (1945)

areas close to the poleward limits of its travel (Musk 1983). In West Africa, for example, if the ITCZ fails to reach its normal northern limits at about 20° N, the regions at that latitude will experience drought. The greatest problems arise when such dry years occur in groups. The population might be able to survive one or two years of drought, but a longer series has cumulative effects similar to those in the 1960s and 1970s, when consecutive years of drought led to famine, malnutrition, disease, and death. In contrast, years in which the ITCZ with its accompanying rain moves farther north than normal, or remains at its poleward limits for a few extra days or even weeks, can be considered as good years.

In the past, this variability was very much part of the way of life of the drought-prone areas in sub-Saharan Africa. The drought of 1968 to 1973 in the Sahel was the third major dry spell to hit the area this century, and, although the years following 1973 were wetter, by 1980 drier conditions had returned (see Figure 3.6). By 1985 parts of the region were again experiencing fully fledged drought (Cross 1985a). Much as the population must have suffered in the past, there was little they could do about it, and few on the outside showed much concern. Like all primitive nomadic populations, the inhabitants of the Sahel increased in good years and decreased in bad, as a result of the checks and balances built into the environment.

That situation has changed somewhat in recent years. The introduction of new scientific medicine, limited though it may have been by western standards, brought with it previously unknown

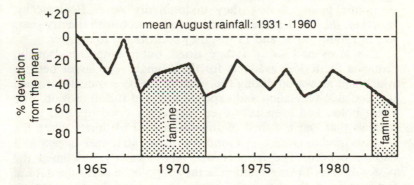

Figure 3.6 August rainfall in the Sahel: 1964–84

Source: After Cross (1985a)

medical services, such as vaccination programmes, and led to improvements in child nutrition and sanitation. Together, these helped to lower the death rate, and, with the birth rate remaining high, populations grew rapidly, doubling between 1950 and 1980 (Crawford 1985). The population of the Sahel grew at a rate of 1 per cent per annum during the 1920s, but by the time of the drought in the late 1960s the rate was as high as 3 per cent (Ware 1977). Similar values have been calculated for the eastern part of the dry belt in Somalia (Swift 1977) and Ethiopia (Mackenzie 1987b). Initially, even such a high rate of growth produced no serious problems, since in the late 1950s and early 1960s a period of heavier more reliable rainfall produced more fodder and allowed more animals to be kept. Crop yields increased also in the arable areas. Other changes, with potentially serious consequences, were taking place at the same time, however. In the southern areas of the Sahel, basic subsistence farming was increasingly replaced by the cash-cropping of such commodities as peanuts and cotton. The way of life of the nomads had already been changed by the establishment of political boundaries in the nineteenth century, and the introduction of cash-cropping further restricted their ability to move as the seasons dictated. Commercialization had been introduced into the nomadic community also, and in some areas market influences encouraged the maintenance of herds larger than the carrying capacity of the land. This was made possible to some extent by the drilling or digging of new wells, but as Ware (1977) has pointed out, the provision of additional water without a parallel provision for additional pasture served only to aggravate ecological problems. All of these changes were seen as improvements when they were introduced, and from a socio-economic point of view they undoubtedly were. Ecologically, however, they were suspect, and in combination with the dry years of the 1960s, 1970s, and 1980s, they contributed to disaster.

The grass and other forage dried out during the drought, reducing the fodder available for the animals. The larger herds – which had grown up during the years of plenty – quickly stripped the available vegetation and exposed the land to soil erosion. The water holes, and eventually even the rivers, dried up, and those animals that had not died of starvation died of thirst. When the nomads tried to move they found that it was no longer as easy as it had once been. The farmers who had formerly welcomed the pastoralists for the meat and milk they supplied and for the natural fertilizer from their animals, no longer wanted herds trampling fields which, with the aid of irrigation, were being cropped all year round. Furthermore, growing peanuts and cotton and only a little food for their own use, they had no excess left for the starving nomads. Eventually, as the drought continued, the farmers suffered also. There was insufficient water for the irrigation systems, the artificial fertilizers that had replaced the animal product was less effective at low moisture levels, and yields were reduced dramatically. In a desperate attempt to maintain their livelihood, they seeded poorer land, which was soon destroyed by soil erosion in much the same way as the overgrazed soil of the north had been.

The net result of such developments was the death of millions of animals – probably five million cattle alone – and several hundred thousand people. The latter numbers would have been higher but for the outside aid which provided for seven million people at the peak of the drought in the Sahel between 1968 and 1973 (Glantz 1977). Similar numbers were involved in Ethiopia between 1983 and 1985, although accurate figures are difficult to obtain because of civil war in the area (Cross 1985b). Drought and locusts destroyed crops in the northern Ethiopian provinces of Tigray and Eritrea again in 1987, and the UN World Food Programme estimated that as many as three million inhabitants were put at risk of starvation and death (Mackenzie 1987a). Elsewhere – in Kenya, Uganda, Sudan, and Mozambique – severe drought continues. Even in the Sahel, which had experienced some improvement in the late 1970s, increasing aridity after 1980 was the precursor of the more intense drought affecting the area once again. The death of the animals, the destruction of the soil and, indeed, the destruction of society, has meant that all of sub-Saharan Africa – from Senegal to Somalia – remains an impoverished region, dependent upon outside aid and overshadowed by the ever present potential for disaster the next time the rains fail, as fail they will.

Drought on the Great Plains

Since all of the nations stricken by drought in sub-Saharan Africa are under-developed, it might be assumed that lack of economic and technological development contributed to the problem. To some extent it did, but it is also quite clear that economic and technological advancement is no guarantee against drought. The net effects may be lessened, but the environmental processes act in essentially the same way, whatever the stage of development.

This is well illustrated in the problems faced by farmers on the Great Plains that make up the interior of North America (see Figure 3.7). Stretching from western Texas in the south, along the flanks of the Rocky Mountains to the Canadian prairie provinces

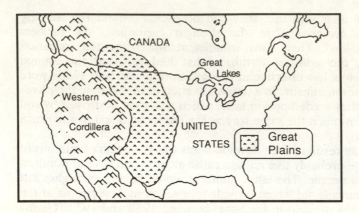

Figure 3.7 The Great Plains of North America

in the north, they form an extensive area of temperate grassland with a semi-arid climate. They owe their aridity in part to low rainfall, but the situation is aggravated by the timing of the precipitation, which falls mainly in the summer months, when high temperatures cause it to be rapidly evaporated. Contingent drought, brought about by the variable and unpredictable nature of the rainfall, is characteristic of the area – consecutive years may have precipitation 50 per cent above normal or 50 per cent below normal – and this has had a major effect on the settlement of the plains. Averages have little real meaning under such conditions, and agricultural planning is next to impossible. The tendency for wet or dry years to run in series introduces further complexity. Strings of dry years during the exploration of the western plains in the nineteenth century, for example, gave rise to the concept of the Great American Desert. Although the concept has been criticized for being as much myth as reality, it is now evident that several expeditions to the western United States in the early part of the century (Lawson and Stockton 1981), and some decades later to western Canada (Spry 1963), encountered drought conditions which are now known to be quite typical of the area, and which have been repeated again and again since that time.

Historical drought

Local drought is not uncommon on the plains. Almost every year in the Canadian west there are areas which experience the limited precipitation and high temperatures necessary for increased aridity

52

(McKay *et al.* 1967), and archaeological evidence suggests that from the earliest human habitation of the area it has been so (Van Royen 1937). The original inhabitants, who were nomadic hunters, probably responded in much the same way as the people of the Sahel when drought threatened. They migrated, following the animals to moister areas. At times even migration was not enough: for example, during the particularly intense Pueblo drought of the thirteenth century the population was much reduced by famine (Fritts 1965). Major drought episodes, extensive in both time and place, are also a recurring feature of the historical climatology of the Great Plains (Bark 1978).

The weather was wetter than normal when the first European agricultural settlers moved into the west in the late 1860s following the Civil War. The image of the Great American Desert had paled, and the settlers farmed just as they had done east of the Mississippi or in the mid-west, ploughing up the prairie to plant wheat or corn. By the 1880s and 1890s, the moist spell had come to an end, and drought once more ravaged the land (Smith 1920), ruining many settlers and forcing them to abandon their farms. Ludlum(1971) has estimated that fully half the settlers in Nebraska and Kansas left the area at that time, like the earlier inhabitants, seeking relief in migration, in this case back to the more humid east. Some of those who stayed experimented with dry-farming, but even that requires a modicum of moisture if it is to be successful. Others allowed the land to revert to its natural state, and used it as grazing for cattle, a use for which it was much more suited in the first place. The lessons of the drought had not been learned well, however. Later, in the 1920s, the rains seemed to have returned to stay and a new generation of arable farmers moved in. Lured by high wheat prices, they turned most of the plains over to the plough, seemingly unconcerned about the previous drought (Watson 1963). Crops were good, as long as precipitation remained above normal. By 1931, however, the good years were all but over, and the drought of the 1930s, in combination with the Depression, created such disruption of the agricultural and social fabric of the region that the effects have reverberated down through the decades. Even today, every dry spell is compared to the benchmark of the Dustbowl. The only possible response for many of the drought victims of the 1930s was migration, as it had been in the past. The Okies of the *Grapes of Wrath*, leaving behind their parched farms, had much in common with the Indians who had experienced the Pueblo drought six centuries before. The societies were quite different, but they felt the same pressures, and responded in much the same way. By migrating, both were making the ultimate adjustment to a hostile environment.

If the drought of the 1930s brought with it hardship and misery, it also produced the final realization that drought on the plains is an integral part of the climate of the area. Intense dry spells have recurred since then – particularly in the mid-1970s, the early 1980s (Phillips 1982), and again in 1987 and 1988 – causing significant disruption at the individual-farm and local level. Because of the general acceptance of the limitations imposed by aridity, however, the overall impact was less than it would have been half a century earlier. New agricultural techniques – involving dry farming as well as irrigation – coupled with a more appropriate use of the land, help to offset the worst effects of the arid environment, but some problems will always remain.

Desertification

Although often severe, the problems which arise in most areas which experience seasonal or contingent drought are seen as transitory, disappearing when the rains return. If the rains do not return, the land becomes progressively more arid until, eventually, desert conditions prevail. This is the process of desertification in its simplest form. When considered in this way, desertification is a natural process which has existed for thousands of years, is reversible, and has caused the world's deserts to expand and contract in the past. This is, however, only one approach to desertification – one which sees the process as the natural expansion of desert or desert-like conditions into an area where they had not previously existed. This process, occurring along the tropical desert margins, was referred to originally as 'desertization' (Le Houérou 1977), but that term has been replaced by 'desertification' and the concept has been expanded to include a human element.

There is no widely accepted definition of desertification. Most modern approaches, however, recognize the combined impact of adverse climatic conditions and the stress created by human activity (Verstraete 1986). Both have been accepted by the United Nations as the elements that must be considered in any working definition of the process (Glantz 1977), although the relative importance of each of these elements remains very controversial. Some see drought as the primary element, with human intervention aggravating the situation to such an extent that the overall expansion of the desert is increased. Others see direct human activities as instigating the process. In reality, there must be many causes (see Figure 3.8) which together bring desert-like conditions to perhaps as much as 60,000 sq. km of the earth's surface every

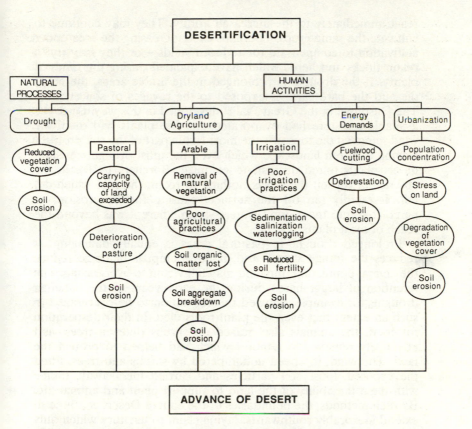

Figure 3.8 The causes and development of desertification

year and threaten up to 30 million sq. km more (see Figure 3.1b). At least 50 million people are directly at risk of losing life or livelihood in these regions. Although these specific numbers are not universally accepted, they are the most frequently quoted. They give an indication of the magnitude of the problem, and are the reason that there has been increasing cause for concern in recent years (van Ypersele and Verstraete 1986).

Desertification initiated by drought

Human nature being what it is, when drought strikes there is a natural tendency to hope that it will be short and of limited intensity. The inhabitants of drought-prone areas, therefore, may not react immediately to the increased aridity. They may continue to cultivate the same crops – perhaps even increasing the area under cultivation to compensate for reduced yields – or they may try to retain flocks and herds which have expanded during the times of plenty. If the drought is prolonged in the arable areas, the crops die and the bare earth is exposed to the ravages of soil erosion. The Dustbowl in the Great Plains developed in this way. Once the available moisture had evaporated and the plants had died, the wind removed the topsoil – the most fertile part of the soil profile – leaving a barren landscape which even the most drought-resistant desert plants found difficult to colonize (Borchert 1950). In the absence of topsoil there was nothing to retain the rain which did fall. It rapidly ran off the surface, causing further erosion, or percolated into the ground-water system where it was beyond the reach of most plants.

Prolonged drought in pastoral areas is equally damaging. It reduces the forage supply and, if no attempt is made to reduce the animal population, the land may fall victim to overgrazing. The retention of larger herds during the early years of the Sahelian drought, for example, allowed the vegetation to be overgrazed to such an extent that even the plant roots died. In their desperation for food, the animals also grazed on shrubs or even trees, and effectively removed vegetation which had helped to protect the land. The wind, its speed unhampered by shrubs and trees, lifted the exposed, loose soil particles and carried them away, taking with them the ability of the land to support plant and animal life. By such methods, the boundaries of the Sahara Desert continue to extend inexorably southwards, laying claim to territory which only recently supported a population living as comfortably as it could within the constraints of the environment.

Desertification caused by human activity

Climatic variability clearly made a major contribution to desertification, in both the Sahel and the Great Plains, perhaps even initiating the process, and in concert with human activities created serious environmental problems. An alternative view sees human activity in itself as being capable of initiating desertification in the absence of increased aridity (Verstraete 1986). For example, human interference in areas where the environmental balance is a delicate one might be sufficient to set in motion a train of events leading eventually to desertification. The introduction of arable agriculture into areas more suited to grazing, or the removal of forest cover to open up agricultural land or to provide wood for fuel, may disturb the ecological balance to such an extent that the quality of the environment begins to decline. The soil takes a physical beating during cultivation: its crumb structure is broken down and its individual constituents are separated from each other. In addition, cultivation destroys the natural humus in the soil and the growing crops remove the nutrients, both of which normally help to bind the soil particles together into aggregates. If nothing is done to replace the organic material or the nutrients, the soil then becomes highly susceptible to erosion. Modern agricultural techniques which allow the soil to lie exposed and unprotected by vegetation for a large part of the growing season, also contribute to the problem. When wind and water erode the topsoil it becomes impossible to cultivate the land, and even natural vegetation has difficulty in re-establishing itself in the shifting mineral soil that remains.

The removal of trees and shrubs to be used as fuel has had similar effects in many Third World nations, where the main source of energy is wood. Le Houérou (1977) has estimated that, in the areas along the desert margins in Africa and the Middle East, a family of five will consume, every year, all of the fuel available on 1 hectare of woody steppe. With a population close to 100 million dependent upon this form of fuel in the area concerned, as much as 25 million hectares per year are being destroyed, and all of that area is potentially open to desertification.

No change in land use is required to initiate such a progression in some regions. The introduction of too many animals into an area may lead to overgrazing and cause such environmental deterioration that after only a few years the land may no longer be able to support the new activity. Forage species are gradually replaced by weeds of little use to the animals, and the soil becomes barren and unable to recover even when grazing ceases. In all of these cases, the land has been laid waste with little active contribution from climate. Human activities have disturbed the environmental balance to such an extent that they have effectively created a desert.

The prevention and reversal of desertification

Faced with the loss of 60,000 sq. km of land to the desert every year – with all the social, economic, political, and environmental problems that such a figure implies – it is inevitable that two questions have received increased attention in recent years. Can desertification be prevented? Can the desertification which has already happened be reversed? The answer to both is a qualified yes (see Table 3.1). The environment itself took care of these in the past, but human interference has ensured that natural checks and balances are now much less effective. However, the starvation and death that has stalked the Sahel and Ethiopia in recent years might be taken as an indication that the environment continues in its attempt to regain the ecological balance necessary for the solution of the problems.

In theory, society could work with the environment by developing a good understanding of environmental relationships in the threatened areas or by assessing the capability of the land to support certain activities and by working within the constraints that these would provide. In practice, non-environmental elements – such as politics and economics – may prevent the most ecologically appropriate use of the land. A typical response to the variable precipitation in areas prone to desertification is to

Table 3.1 Action required for the prevention and reversal of desertification

Prevention	Reversal
(a) Good land-use planning and management: e.g. cultivation only where and when precipitation is adequate; animal population based on carrying capacity of land in driest years; maintenance of woodland where possible.	(a) Prevention of further soil erosion: e.g. by contour ploughing; by gully infilling; by planting or constructing windbreaks.
(b) Irrigation appropriately managed to minimize sedimentation, salinization, and waterlogging.	(b) Reforestation.
(c) Plant breeding for increased drought resistance.	(c) Improved water use: e.g. storage of run-off; well-managed irrigation.
(d) Improved long-range drought forecasting.	(d) Stabilization of moving sand: e.g. using matting; by re-establishment of plant cover; using oil waste mulches and polymer coating.
(e) Weather modification: e.g. rainmaking; snowpack augmentation.	(e) Social, cultural, and economic controls: e.g. reduction of grazing animal herd size; population resettlement.
(f) Social, cultural, and economic controls: e.g. population planning; planned region economic development; education.	

consider the good years as normal and to extend production into marginal areas at that time (Riefler 1978). The stage is then set for progressive desertification when the bad years return. Experience in the United States has shown that this can be prevented by good land-use planning, which includes not only consideration of the best use of the land, but also the carrying capacity of that land under a particular use (Sanders 1986). To be effective in an area such as Africa, this would involve restrictions on grazing and culti-vation in many regions, but not only that. Estimates of the carrying capacity of the land would have to be based on conditions in the worst years rather than in the good or even normal years (Kellogg and Schneider 1977; Mackenzie 1987b). Such actions would undoubtedly bring some improvement to the situation, but for the inhabitants of the area the transition between old and new systems might well be highly traumatic. The pastoralists would lose the advantages of larger flocks and herds in the good years, some of the cultivators might have to allow arable land to revert to pasture – or reduce cash-cropping and return to subsistence agriculture – and members of both groups might have to give up their rural life-style and become urbanized.

The problem of the destruction of woodland will also have to be addressed if desertification is to be prevented. Trees and shrubs protect the land against erosion, yet they are being cleared at an alarming rate. One hundred years ago in Ethiopia, 40 per cent of the land could be classified as wooded; today only 3 per cent can be designated in that way (Mackenzie 1987b). Good land-use planning would recognize that certain areas are best left as wood-land, and would prevent the clearing of that land for the expansion of cultivation or the provision of fuelwood. The latter problem is particularly serious in most of sub-Saharan Africa where wood is the only source of energy for most of the inhabitants. It has ramifications which reach beyond fuel supply. Experience has shown that where wood is not available, animal dung is burned as fuel, and although that may supply the energy required, it also represents a loss of nutrients which would normally have been returned to the soil. Any planning involving the conservation of fuelwood must consider these factors, and make provision for an alternative supply of energy or another source of fertilizer.

Many of the techniques which could be employed to prevent desertification are also considered capable of reversing the process. Certainly there are areas where the destruction of the land is now irreversible, but there have also been some successes. In parts of North America, land apparently destroyed in the 1930s has been successfully rehabilitated through land-use planning and direct soil-conservation techniques such as contour ploughing, strip cropping, and the provision of windbreaks. Irrigation has also become common, and methods of weather modification, mainly rain making, have been attempted, although with inconclusive results (Rosenberg 1978). Many of these methods could be applied with little modification in areas such as Africa, where desertification is rampant. Dry-farming techniques have been introduced into the Sudan (CIDA 1985); in Ethiopia, new forms of cultivation similar to contour ploughing have been developed to conserve water and prevent erosion (Cross 1985b); in Mali and other parts of West Africa, reforestation is being attempted to try to stem the southward creep of the desert (CIDA 1985). The lack of moisture has been tackled directly in many areas by the drilling of boreholes to produce groundwater, although without strict control this may not be the best approach. Extra water encourages larger flocks and herds which overgraze the area around the borehole. Le Houérou (1977) has pointed out that the pasture has been completely destroyed for 15–30 km around some of the boreholes drilled at the time of the Sahelian drought.

All of these developments deal directly with the physical symptoms of desertification, but it has been argued that many studies have overlooked economic and social constraints (Hekstra and Liverman 1986). Ware (1977) has suggested, for example, that insufficient development of markets, transportaion, and welfare systems made a major contribution to the problems in the Sahel, and future planning must give these factors due consideration. Similarly, population growth rates and densities must be examined with a view to reducing human pressure on the land. Relief may come in the form of family planning or through relocation, and despite potentially serious social and political concerns, this may be the only way to tackle that aspect of the problem (Mackenzie 1987b).

Drought prediction

Even if all of these methods of dealing with drought, famine, and desertification were to be initiated immediately, the results would be a long time coming. In Africa this means that attention to the existing and recurring problems of drought and famine must continue. To be effective, such aid requires: an early warning of the problem; fast response; and timely delivery of relief. The last two are essentially socio-economic elements, but the first is physical, and it has led to numerous attempts by climatologists in recent years to devise a method by which drought may be predicted.

Drought is not the sole cause of famine or desertification, yet it is certainly a major cause – often initiating the problem only to have it intensified by other factors. Prevention of drought is not feasible at present, nor would it necessarily bring about an end to famine and desertification if it was. If drought could be predicted, however, responses could be planned, and the consequences therefore much reduced. The simplest approach is the actuarial forecast, which estimates the probability of future drought based on past occurrences. To be successful, actuarial forecasting requires a lengthy sequence of data for analysis (Schneider 1978). In many areas, including sub-Saharan Africa, the record is simply too short to provide a reliable prediction. Problems with the homogeneity of the meteorological record may also reduce the significance of the results.

An extension of the actuarial approach is the linking of the meteorological variables with some other environmental variable which includes a recognized periodicity in its behaviour (Oguntoyinbo 1986). One of the most commonly cited links of this type is the relationship between sun-spot activity and precipitation (see Figure 3.9). In North America, drought on the plains has been correlated with the minimum of the 22-year double sun-spot cycle. The drought years of the mid-1970s, for example, coincided with a period of minimum sun-spot activity. The previous drought, some 20 years earlier in the mid-1950s, also fitted into the cycle. Close as such a correlation may seem, it applies less well outside the

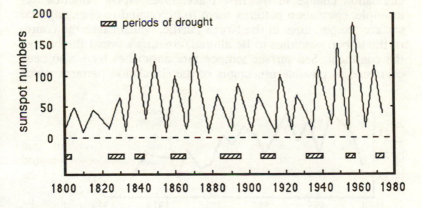

Figure 3.9 A comparison of drought and sun-spot cycles in western North America

Source: Compiled from data in Schneider and Mass (1975), Schneider and Mesirow (1976)

61

western United States. Furthermore, the relationship remains a statistical one, and, as Schneider (1978) has pointed out, there is no physical theory to explain the connection between the two phenomena.

In the search for improved techniques of drought prediction, much time and effort have gone into the study of the physical causes of drought. The immediate causes commonly involve changes in atmospheric circulation patterns. Drought in the western prairies of Canada and the United States, for example, is promoted by a strong zonal airflow, which brings mild Pacific air across the western mountains. As it flows down the eastern slopes, it warms up, and its relative humidity decreases, causing the mild dry conditions in winter and the hot, dry conditions in summer, which produce drought (Sweeney 1985). Seasonal drought in the Sahel was long linked to the failure of the ITCZ to move as far north as normal during the northern summer (see Figure 3.10). This is no longer generally accepted as sufficient to explain the lengthier dry spells, however, and the Sahelian drought is now being examined as part of the broader pattern of continent-wide rainfall variability, associated with large-scale variations in the atmospheric circulation (Nicholson 1989).

This type of knowledge is, in itself, of limited help in predicting or coping with drought, since the drought has usually arrived by the time the characteristic circulation patterns are recognized. It is necessary to move back a stage to try to find out what caused the circulation change in the first place. Over North America, for example, circulation patterns seem to be related to changing sea-surface temperatures in the North Pacific, which cause the course of the upper westerlies to be altered, creating a zonal flow across the continent. Sea-surface temperature anomalies have also been examined as possible precursors of the circulation patterns which

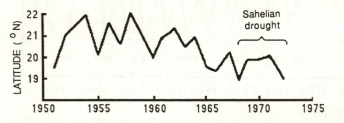

Figure 3.10 Variations in the northward penetration of the monsoon rains in the Sahel, 1950–72

Source: After Bryson and Murray (1977)

cause drought in the Sahel, but the results remain inconclusive (Folland *et al.* 1986, Owen and Ward 1989).

It has also been suggested that human activities have enhanced the physical processes responsible for the drought. Charney (1975) proposed that overgrazing and wood-cutting in the Sahel increased the surface albedo and disrupted the regional radiation balance. Surface heating declined as more solar radiation was reflected, and this, in turn, caused some cooling of the atmosphere. This cooling encouraged subsidence, or augmented existing subsidence, and helped to reduce the likelihood of precipitation by retarding convective activity. With less precipitation, vegetation cover decreased and the albedo of the surface was further enhanced. This process was described as a bio-geophysical feedback mechanism, and, although it may not fit the situation in the Sahel exactly (Courel *et al.* 1984), recent circulation-model experiments show that it is a theory worthy of further analysis, using surface data collected in the area (Laval 1986).

While such studies may provide a better understanding of the problems of drought and the mechanisms involved, they are insufficient to provide a direct forecasting mechanism. Researchers have re-examined certain relationships in the earth/atmosphere system in search of something more suitable. Their approach is based on the observation that the various units in the system are interconnected in such a way that changes in one unit will automatically set in motion changes in others. Since many of the changes are time-lagged, it should be possible to predict subsequent developments if the original change can be recognized. This forms the basis of the concept of teleconnection, or the linking of environmental events in time and place. In recent years, the search for a drought-forecasting mechanism involving teleconnection has centred on two related elements. The first of these is the Southern Oscillation, a periodic fluctuation in atmospheric pressure in the southern Pacific, first recognized in the 1920s by Sir George Walker as he sought to develop methods for forecasting the rainfall in the Indian monsoon. Intimately linked with this oscillation is El Niño, a flow of anomalously warm surface water which appears with some regularity in the equatorial regions of the eastern Pacific. The name originally referred to a warm current which appeared off the coast of Peru close to Christmas – hence El Niño, the (Christ) Child – but now it is applied to a larger scale phenomenon (Lockwood 1984). The term ENSO is commonly used to refer to the combination of El Niño and the Southern Oscillation.

An indication of the Southern Oscillation is obtained by

comparing barometric pressure differences between Tahiti, in the eastern Pacific, and Darwin, in northern Australia. Pressure at these two stations is negatively correlated, high pressure over Tahiti normally being accompanied by low pressure over Darwin, for example. In contrast, low pressure at Tahiti is matched by high pressure at Darwin. It is this reversal of the regional pressure patterns which is referred to as the Southern Oscillation (see Figure 3.11). It has a periodicity of 1–5 years, and in its wake it brings changes in wind fields, sea-surface temperatures, and ocean-circulation patterns. When pressure is high over Tahiti and low over Darwin, the general wind and surface-water flow is from east to west, and there is a tendency for relatively warm water to pond up at the western end of the Pacific Ocean. The removal of the warm surface water from the eastern Pacific allows the upwelling of relatively cold water from below in that area. These conditions are reversed following the Oscillation. Easterly winds are replaced by the westerlies, as the atmospheric pressure changes, and warm water begins to flow east again to replace the cold (see Figure 3.12). It is this phenomenon which has come to be called El Niño. When it is strongly developed, it may keep areas in the eastern Pacific warmer than normal for periods of up to a year (Rasmusson and Hall 1983).

The large shifts of air and water associated with these developments cause major alterations to energy-distribution patterns. Zonal energy flow replaces the meridianal flow which is normal in tropical Hadley Cells, and, because of the integrated nature of the atmospheric circulation, the effects are eventually felt beyond the tropics. The impact of all of this on drought can be seen in the tele-

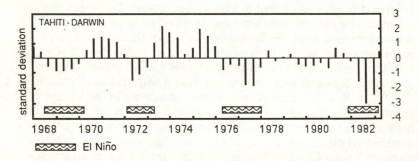

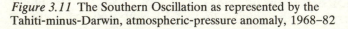

Figure 3.11 The Southern Oscillation as represented by the Tahiti-minus-Darwin, atmospheric-pressure anomaly, 1968–82

Source: Adapted from Rasmusson and Hall (1983)

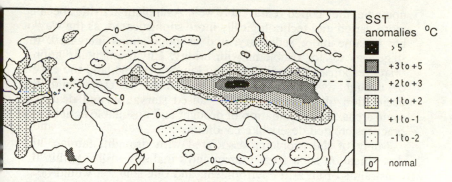

SST
anomalies °C

▓ > 5

▓ +3 to +5

▓ +2 to +3

▓ +1 to +2

☐ +1 to -1

☐ -1 to -2

☐ normal

Figure 3.12 Sea-surface temperature (SST) anomalies in the Pacific
Ocean, December 1982

Source: After Rasmusson and Hall (1983)

connections which have been established between ENSO events
and precipitation in areas as far apart as Brazil, India, Indonesia,
and Australia. Drought in north-eastern Brazil commonly occurs
in conjunction with an El Niño event, and India receives less
monsoon rainfall during El Niño years. There is, however, no clear
relationship between drought in the Sahel and the occurrence of
ENSO events (Lockwood 1986).

Teleconnection links remain largely theoretical, although most
of these relationships can be shown to be statistically significant.
Certainly, in the study of drought, it has not been possible to
establish physical relationships which would allow increasing
aridity to be predicted with any accuracy. However, it seems
probable that future developments in drought prediction are likely
to include consideration of teleconnections, particularly those
involving time-lags, established in conjunction with the improvement of general circulation models (Oguntoyinbo 1986).

Summary

In many parts of the world, low precipitation levels combine with
high evapotranspiration rates to produce an environment characterized by its aridity. Under natural conditions, the ecological
elements in such areas are in balance with each other and with the
low moisture levels. If these change, there is a wholesale readjust-
ment as the environment attempts to attain balance again. The
early inhabitants of these areas also had to respond to the changes,
and, as long as their numbers remained small and their way of life

nomadic, they coped remarkably well. As human activities became more varied and technologically more sophisticated, as the way of life became more sedentary, and as populations increased, the stage was set for problems with aridity. Environmentally appropriate responses to aridity were no longer possible, and the effects of drought were magnified. The failure of crops and the decimation of flocks and herds caused starvation and death. In some areas, the combination of drought and unsuitable agricultural practices created desert-like conditions.

Many of the problems associated with drought, famine, and desertification stem from humankind's inability to live within the constraints of an arid environment, and one solution would be to restrict activities in drought-prone regions. Given existing political, cultural, and socio-economic realities, such an approach is not feasible in most of the affected areas. Unfortunately, many of the alternative solutions are short-term in their impact – of necessity in many cases – and, in some areas at least, may be setting up even greater difficulties in years to come. In short, solutions to the problems of drought, famine, and desertification are unlikely to be widely available in the forseeable future, and the images most recently generated in Ethiopia are likely to recur with sickening frequency.

Suggestions for further study

1   Examine the reports of such aid organizations as Oxfam, Save the Children, CARE, and UNICEF. How do they contribute to the solution of the problems of drought, famine, and desertification? Are their approaches to the problems always environmentally sound, or is there some validity to the suggestion that they treat the symptoms rather than the disease?

2. Using the data provided:

    (a) Draw a graph for each station to show precipitation and temperature distribution throughout the year. Be neat and include all the information usually required on graphs: e.g. title, axis labels, key.

    (b) For each station attempt to correlate the annual patterns of temperature and precipitation with the various zones (A, B, C, D) associated with the Inter-tropical Convergence. Indicate these zones on the graphs.

emperature and precipitation statistics for selected stations in West Africa

	J	F	M	A	M	J	J	A	S	O	N	D	Year	
imbuktu Mali)	22	24	29	32	35	35	33	31	32	32	27	22	30	T °C
6° 49′ N	0	0	2.5	0	5.0	23	68	68	25	2.5	0	0	194	Pmm
imbo Guinea)	22	25	27	27	25	23	22	22	22	23	22	22	24	T °C
)° 40′ N	0	0	25	60	160	225	310	368	255	168	33	0	1,604	Pmm
adan Nigeria)	27	28	29	28	27	26	25	25	25	26	27	27	26	T °C
23′ N	10	23	90	133	148	185	153	83	175	155	45	10	1,210	Pmm
kassa Nigeria)	26	26	27	27	26	25	25	25	25	25	26	26	26	T °C
19′ N	65	163	250	215	425	465	253	233	483	618	265	163	3,598	Pmm

Chapter four

Acid rain

The unrelenting pollution of the atmosphere by modern society is at the root of several global environmental issues. The emission of pollutants into the atmosphere is one of the oldest human activities, and the present problems are only the most recent in a lengthy continuum. Issues such as acid rain, increased atmospheric turbidity, and the depletion of the ozone layer include essentially the same processes which led to the urban air-pollution episodes of a decade or two ago. The main difference is one of scale. The previous problems had a local, or at most, a regional impact. The current issues are global in scope, and therefore potentially more threatening.

The nature and development of acid rain

Acid rain is normally considered to be a by-product of modern atmospheric pollution. Even in a pure, uncontaminated world, however, it is likely that the rainfall would be acidic. The absorption of carbon dioxide by atmospheric water produces weak carbonic acid, and nitric acid may be created during thunderstorms, which provide sufficient energy for the synthesis of oxides of nitrogen (NO_x) from atmospheric oxygen and nitrogen. During volcanic eruptions or forest fires, sulphur dioxide (SO_2) is released into the atmosphere to provide the essential component for the creation of sulphuric acid. Acids formed in this way fall to earth in rain and that process can be considered as one of the atmosphere's self-cleansing mechanisms. The acids become involved in a variety of physical and biological processes once they reach the earth's surface. The return of nitrogen and sulphur to the soil in this way helps to maintain nutrient levels, for example. The peculiar landscapes of limestone areas – characterized by highly weathered bedrock, rivers flowing in steep-sided gorges or through inter-connected systems of underground-stream channels and caves – provide excellent examples of what even moderately acid rain can do.

In reality, since 'acid rain' includes snow, hail, and fog as well as rain, it would be more appropriate to describe it as 'acid precipitation'. The term 'acid rain' is most commonly used for all types of 'wet deposition', however. A related process is 'dry deposition', which involves the fall-out of the oxides of sulphur and nitrogen from the atmosphere, either as dry gases or adsorbed on other aerosols such as soot or fly-ash (Park 1987). These particles may not be in an acidic state while in the atmosphere, but on contact with moisture in the form of fog, dew, or surface water, for example, they become acidic, and have similar effects as the constituents of wet deposition. At present, both wet and dry deposition are normally included in the term 'acid rain' and, to maintain continuity, that convention will be followed here.

Current concern over acid rain is not with the naturally produced variety, but rather with that which results from modern industrial activity. Technological advancement in a society often depends upon the availability of metallic ores, which can be smelted to produce the great volume and variety of metals needed for industrial and socio-economic development. As a by-product of the smelting process, particularly when non-ferrous ores are involved, considerable amounts of SO_2 are released into the atmosphere. The burning of coal and oil, to provide energy for space heating or to fuel thermal electric power stations, also produces SO_2. The continuing growth of transportation systems using the internal combustion engine, another characteristic of a modern technological society, contributes to acid rain through the release of NO_x into the atmosphere.

Initially, the effects of these pollutants were restricted to the local areas in which they originated, and where their impact was often obvious. For example, the detrimental effects of SO_2 on vegetation around the smelters at Sudbury (Ontario), Trail (British Columbia), Anaconda (Montana), and Sheffield (England) have long been recognized (Garnett 1967, Hepting 1971). As emissions increased, and the gases were gradually incorporated into the larger-scale, atmospheric circulation, the stage was set for an intensification of the problem. Sulphur compounds of anthropogenic origin are now blamed for as much as 65 per cent of the acid rain in eastern North America, with nitrogen compounds accounting for the remainder (Ontario: Ministry of the Environment 1980). In Europe, emission totals for SO_2 and NO_x suggest that there the split is closer to 75 per cent and 25 per cent (Park 1987).

Acid precipitation produced by human activities differs from natural acid precipitation not only in its origins, but also in its quality; anthropogenically produced acid rain tends to be many times more acidic than the natural variety. The acidity of a solution is indicated by its hydrogen ion concentration or pH (potential hydrogen); the lower the pH, the higher the acidity. A chemically neutral solution, such as distilled water, has a value of 7, with increasingly alkaline solutions ranging from 7 to 14, and increasingly acidic solutions ranging from 7 down to 0 (see Figure 4.1). This pH scale is logarithmic, and, as a result, a change of one point represents a tenfold increase or decrease in the hydrogen ion concentration, while a two-point change represents a one hundred-fold increase or decrease. A solution with a pH of 4.0 is ten times more acidic than one of pH 5.0; a solution of pH 3.0 is one hundred times more acidic than one of pH 5.0.

The difference between 'normal' and 'acid' rain is commonly of the order of 1.0 to 1.5 points. In North America, for example,

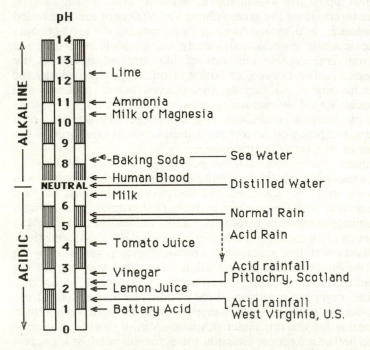

Figure 4.1 The pH scale, showing the pH level of acid rain in comparison to that of other common substances

naturally acid rain has a pH of about 5.6, while measurements of rain falling in southern Ontario, Canada, frequently provide values in the range of 4.5 to 4.0 (Ontario: Ministry of the Environment 1980). To put these values in perspective it should be noted that vinegar has a pH of 2.7 and milk a pH of 6.6 (see Figure 4.1). Thus, Ontario rain is about 100 times more acidic than milk, but 100 times less acidic than vinegar. Similar values for background levels of acidic rain are indicated by studies in Europe, although the Central Electricity Generating Board (CEGB) in Britain has argued for pH 5.0 as the normal level for naturally acid rain (CEGB 1984). Remarkably high levels of acidity have been recorded on a number of occasions on both sides of the Atlantic. In April 1974, for example, rain falling at Pitlochry, Scotland, had a pH measured at 2.4 (Last and Nicholson 1982), and a value of 2.7 was reported from western Norway a few weeks later (Sage 1980). North American records also include acid rainfall of pH 2.7, plus a value of pH 1.5, some 11,000 times more acid than normal, for rain falling in West Virginia in 1979 (LaBastille 1981).

The quality of the rain is determined by a series of chemical processes set in motion when acidic materials are released into the atmosphere. Some of the SO_2 and NO_x emitted will return to the surface quite quickly, and close to their source, as dry deposition. The remainder will be carried up into the atmosphere, to be converted into sulphuric and nitric acid, which will eventually return to earth as acid rain. Several variables – including the concentration of heavy metals, the presence of ammonia, the intensity of sunlight, and the humidity of the atmosphere – have an effect on the rate at which the conversion takes place, but the processes involved are fundamentally simple (see Figure 4.2). Oxidation converts the gases into acids, and subsequent dissolution causes the acids to dissociate into a solution of electrically charged particles, called ions. For example, sulphuric acid in solution is a mixture of positively charged hydrogen ions (cations) and negatively charged sulphate ions (anions). It is these solutions, or 'cocktails of ions' as Park (1987) calls them, that constitute acid rain.

Whatever the complexities involved in the formation of acid rain, the time-scale is crucial. The longer the original emissions remain in the atmosphere, the more likely it is that the reactions will be completed, and the sulphuric and nitric acids produced. Long-Range Transportation of Atmospheric Pollution (LRTAP) is one of the mechanisms by which this is accomplished.

Air pollution remained mainly a local problem in the past. The effects were greatest in the immediate vicinity of the sources, and much of the effort of environmental groups in the 1960s and

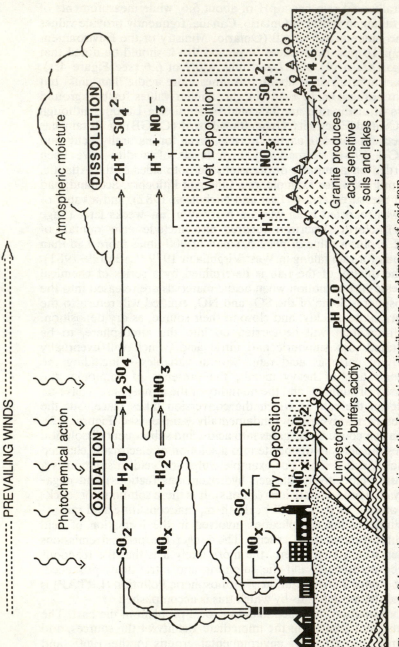

Figure 4.2 Schematic representation of the formation, distribution, and impact of acid rain

Source: Compiled from information in Park (1987); Miller (1984); LaBastille (1981)

1970s was expended in attempts to change that situation. Unfortunately, some of the changes inadvertently contributed to the problem of acid rain. One such was the tall-stacks policy. In an attempt to achieve the reduction in ground-level pollution required by the Clean Air Acts, the CEGB in Britain erected 200-m high smokestacks at its generating stations (Pearce 1982d). Industrial plants and power stations in the United States took a similar approach, increasing the heights of their stacks until, by 1981, at least twenty were more than 300 m high (LaBastille 1981). The International Nickel Company (INCO) added a 400-m superstack to its nickel-smelter complex at Sudbury, Ontario, in 1972 (Sage 1980). The introduction of these taller smokestacks on smelters and thermal electric power stations, along with the higher exit velocities of the emissions, allowed the pollutants to be pushed higher into the atmosphere. This effectively reduced local pollution concentrations, but caused the pollutants to remain in the atmosphere for longer periods of time, thus increasing the probability that the acid-conversion processes would be completed. The release of pollutants at greater altitudes also placed them outside the boundary-layer circulation and into the larger-scale, atmospheric-circulation system with its potential for much greater dispersal through the mechanisms of LRTAP. The net result was a significant increase in the geographical extent of the problem of acid rain.

The geography of acid rain

The main sources of the ingredients of acid rain are to be found in the industrialized areas of the northern hemisphere. North-eastern North America, Britain, and western Europe have received most attention (see Figure 4.3), but eastern Europe and the USSR are also important sources. As early as 1965, air pollution was killing oak and pine trees on Leo Tolstoy's historic country estate at Yosnaya Polyana (Goldman 1971), and the Soviet Union remains the world's leading producer of SO_2 (Pearce 1982b). In Asia, Japanese industries emit large quantities of SO_2 (Park 1987), while the industrial areas of China are also likely to be major contributors of these gases, although statistical information is not readily available from there.

The geographical distribution of acid rain is largely restricted to the industrialized nations of the northern hemisphere at present, but it has the potential to expand to a near-global scale in the future. Much will depend upon the rate at which Third World countries industrialize, and the nature of that industrialization.

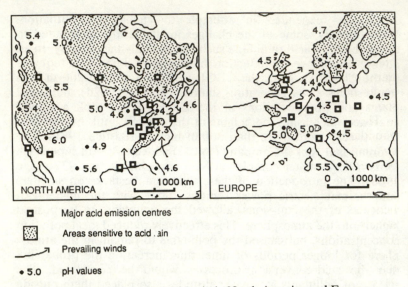

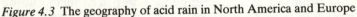

☐   Major acid emission centres

▨   Areas sensitive to acid . ain

╱   Prevailing winds

• 5.0   pH values

Figure 4.3 The geography of acid rain in North America and Europe

Source: Compiled from data in Park (1987); Miller (1984); LaBastille (1981); Ontario, Ministry of the Environment (1980)

Experience in the developed world shows that other possibilities exist also. For example, large conurbations – such as Los Angeles – with few sulphur producing industries but with large volumes of vehicular traffic, have been identified as sources of acidic pollution (Ellis *et al.* 1984). Many developing nations are becoming rapidly urbanized, and, as a result, may provide increased quantities of the ingredients of acid rain in the future (Pearce 1982c).

All of the areas presently producing large amounts of acidic pollution lie within the mid-latitude westerly wind belt. Emissions from industrial activity are therefore normally carried eastwards, or perhaps north-eastwards – often for several hundred kilometres – before being re-deposited. The distance and rate of travel are closely linked to the height of emission. Pollution introduced into the upper westerlies or jet streams is taken further, and kept aloft longer, than that emitted into the boundary-layer circulation. For example, a parcel of air tracked from Toronto, at an altitude of 5000 m, was found well out over the Atlantic, some 950 km east of its source, in less than 12 hours. In the same time span, a parcel of air close to the surface covered less than a quarter of that distance (Cho *et al.* 1984). In this way pollutants originating in the

74

US Midwest cause acid rain in Ontario, Quebec, and the New England states; emissions from the smelters and power stations of the English Midlands and the Ruhr contribute to the acidity of precipitation in Scandinavia. The ultimate example of long-range transportation of atmospheric pollution (LRTAP) is provided by acidic pollution in the Arctic, which has been traced to sources some 8000 km away in North America and Europe (Shaw 1980).

The problem of acid rain obviously transcends national boundaries, introducing political overtones to the problem, and creating the need for international co-operation, if a solution is to be found. That co-operation has not been forthcoming. Since the ingredients of acid pollution are invisible, and the distances they are carried are so great, it is not possible to establish a visible link between the sources of the rain and the areas which suffer its effects. It was therefore easy for polluters to deny fault in the past. The introduction of airborne sampling systems, using balloons (LaBastille 1981), or aircraft (Pearce 1982d) has helped to change that. Experiments with tracer elements added to a polluted airstream (Fowler and Barr 1984) suggest that it may soon be possible to follow emissions from a specific source until they are deposited in acid rain. As monitoring techniques improve, denial of guilt becomes more and more difficult.

Acid rain and geology

The impact of acid rain on the environment depends not only on the level of acidity in the rain, but also on the nature of the environment itself. Areas underlain by granitic or quartzitic bedrock, for example, are particularly susceptible to damage, since the soils and water are already acidic and lack the ability to 'buffer' or neutralize additional acidity from the precipitation. Acid levels therefore rise, the environmental balance is disturbed, and serious ecological damage is the inevitable result. In contrast, areas which are geologically basic, underlain by limestone or chalk, for example, are much less sensitive and may even benefit from the additional acidity. The highly alkaline soils and water of these areas ensure that the acid added to the environment by the rain is very effectively neutralized. In theory, it is important to establish background levels of acidity or alkalinity, so that the vulnerability of the environment to acidification can be estimated; in reality, this is seldom possible, since, in most cases, environmental conditions had already been altered by acid rain, by the time monitoring was introduced.

The areas at greatest risk from acid rain in the northern

75

hemisphere are the pre-Cambrian Shield areas of Canada and Scandinavia, where the acidity of the rocks is reflected in highly acidic
soils and water. The folded mountain structures of eastern Canada,
the United States, Scotland, Germany, and Norway are also
vulnerable (see Figure 4.3). Most of these areas have already
suffered, but the potential for further damage is high. Should the
present emission levels of SO_2 and NO_x be maintained for the next
10–20 years, it is likely that susceptible areas which are presently
little affected by acid rain – in western North America and the
Arctic, for example – would also suffer damage as the level of
atmospheric acidity rises.

Acid rain and the aquatic environment

Park (1987) has summarized the historical development of interest
in acid rain. The earliest concerns were expressed by Robert Smith,
in England, as long ago as 1852, but modern interest in the
problem dates only from the 1960s. Initial attention concentrated
on the impact of acid rain on the aquatic environment, which can
be particularly sensitive to even moderate increases in acidity, and
it was in the lakes and streams on both sides of the Atlantic that
the effects were first apparent. A majority of scientists now accept
that a link exists between the emissions of SO_2 and NO_x and the
acidification of lakes. A minority remains unconvinced. Along
with politicians in the United States and Britain, they continue to
counter requests for emission reductions with calls for more
studies, despite a recent estimate that there have already been
more than 3,000 studies in North America alone (Israelson 1987).
Some uncertainties over the relationship between industrial
emissions and acid rain will always remain, because of the
complexity of the environment, and the variety of its possible
responses to any input. There can be no doubt, however, that acid
rain does fall, and when it does its effects on the environment are
often detrimental.

Acidic lakes are characterized by low levels of calcium and
magnesium and elevated sulphate levels. They also have above-
normal concentrations of potentially toxic metals such as
aluminium (Brakke *et al.* 1988). The initial effect of continued
acid loading varies from lake to lake, since water-bodies differ in
their sensitivity to such inputs. Harmful effects will begin to be felt
by most water-bodies when their pH falls to 5.3 (Henriksen and
Brakke 1988), although damage to aquatic ecosystems will occur
in some lakes before that level is reached, and some authorities
consider pH 6.0 as a more appropriate value (Park 1987). What-

ever the value, once the critical pH level has been surpassed, the net effect will be the gradual destruction of the biological communities in the ecosystem.

There is clear evidence from investigations in areas as far apart as New York State, Nova Scotia, Norway, and Sweden that increased surface-water acidity has adverse effects on fish (Baker and Schofield 1985). The processes involved are complex and their effectiveness varies from species to species (see Figure 4.4). For example, direct exposure to acid water may damage some species. Brook trout and rainbow trout cannot tolerate pH levels much below 6.0 (Ontario: Ministry of the Environment 1980), and at 5.5 smallmouth bass succumb (LaBastille 1981). The *salmonid* group of fish is much less tolerant than coarser fish such as pike

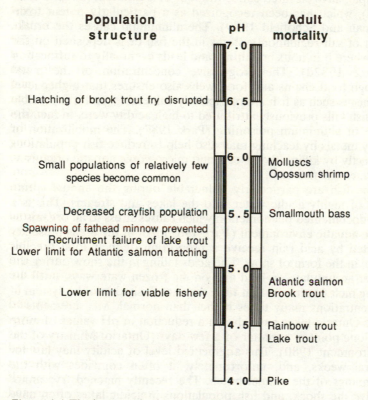

Figure 4.4 The impact of acid rain on aquatic organisms

Source: Compiled from data in Israelson (1987); Baker and Schofield (1985); LaBastille (1981); Ontario, Ministry of the Environment (1980)

77

and perch (Ontario: Ministry of the Environment 1980). Thus, as lakes become progressively more acid, the composition of the fish population changes.

The stage of development of the organism is also important. Adult fish, for example, may be able to survive relatively low pH values, but newly hatched fry, or even the spawn itself, may be much less tolerant (Ontario: Ministry of the Environment 1980). As a result, the fish population in acid lakes is usually wiped out by low reproductive rates even before the pH reaches levels which would kill mature fish (Jensen and Snekvik 1971).

Fish in acid lakes also succumb to toxic concentrations of metals such as aluminium, mercury, manganese, zinc, and lead – leached from the surrounding rocks by the acids. Many acid lakes, for example, have elevated concentrations of aluminium (Brakke *et al.* 1988), which has been recognized as a particularly potent toxin (Cronan and Schofield 1979). The aluminium causes the breakdown of salt-regulation systems in the fish or is deposited on the gills where it inhibits breathing, and leads eventually to suffocation (Pearce 1982d). The progressive concentration of the metal through food chains and food webs also ensures that higher-level organisms such as fish are particularly vulnerable, and it is possible that fish kills previously attributed to high acidity were, in fact, the result of aluminium poisoning (Park 1987). The mobilization of heavy metals by leaching may also help to reduce fish populations indirectly by killing the insects and microscopic aquatic organisms on which the fish feed.

The fish are particularly vulnerable during the annual spring flush of highly acidic water into the lakes and streams. This is a well-documented phenomenon which causes stress to all organisms in the aquatic environment (Park 1987). All of the areas presently affected by acid rain receive a proportion of their total precipitation in the form of snow. The acids falling in the snow during the winter accumulate on land and on the frozen waterways, until the spring melt occurs. At that time they are flushed into the system in concentrations many times higher than normal. Measurements in some Ontario lakes have shown a reduction in pH values of more than one point in a matter of a few days (Ontario: Ministry of the Environment 1980). This augmented level of acidity may last for several weeks, and, unfortunately, it often coincides with the beginnings of the annual hatch. The recently hatched fry cannot survive the shock, and fish populations in acidic lakes often have reduced or missing age groups which reflect this high mortality (Baker and Schofield 1985). The aquatic ecosystems of lakes which are normally well-buffered are not immune to major damage if the melt is rapid and highly acidic (Ontario: Ministry of the Environment 1980).

Fish populations in many rivers and lakes in eastern North America, Britain, and Scandinavia have declined noticeably in the last two to three decades as a consequence of the effects of acid rain. More than 140 lakes in Ontario, Canada, mainly in the vicinity of Sudbury, are fishless and many others have experienced a reduction in the variety of species they contain (Ontario: Ministry of the Environment 1980); several rivers in Nova Scotia, once famous for their Atlantic salmon, are now too acidic to support that fish (Israelson 1987). Damage to fish populations has occurred in an area of 33,000 sq. km in southern Norway, with brown trout particularly hard hit (Baker and Schofield 1985). In nearby Sweden, it is now estimated that some 4000 lakes are fishless (LaBastille 1981). In Britain, no obvious problems emerged until the 1970s, when rivers and lakes in south-west Scotland, the English Lake District, and Wales showed the first signs of decline in the numbers of trout, salmon, and other game fish (Park 1987).

Water-bodies which have lost, or are in the process of losing their fish populations are usually described as 'dead' or 'dying'. This is not strictly correct. All aquatic flora and fauna will decline in number and variety during progressive acidification, but, even at pH 3.5, water boatmen and whirligig beetles survive and multiply (LaBastille 1981), and species of protozoans are found at pH levels as low as 2.0 (Hendrey 1985). Phytoplankton will disappear when pH falls below 5.8 (Almer *et al.* 1974), but acid-tolerant *Sphagnum* mosses will colonize the lake bottoms (Pearce 1982d). Leaves or twigs falling into the water will be slow to decompose, because the bacteria which would normally promote decay have been killed by the acidic conditions. The absence of phytoplankton and the general reduction in organic activity allows greater light penetration, which makes acid lakes unnaturally clear and bluish in colour (LaBastille 1981). This ethereal appearance may suggest death, but even the most acid lakes have some life in them.

Acid rain and the terrestrial environment

Terrestrial ecosystems take much longer to show the effects of acid rain than aquatic ecosystems. As a result, the nature and magnitude of the impact of acid preciptation on the terrestrial environment has been recognized only recently. There is growing evidence that those areas in which the water-bodies have already succumbed to acidification must also face the effects of increasing acid stress on their forests and soils (see Figure 4.5). The threat is

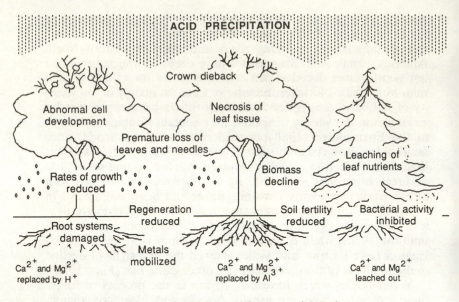

Figure 4.5 The impact of acid rain on the terrestrial environment

Source: Compiled from data in Fernandez (1985); Shriner and Johnston (1985); Tomlinson (1985)

not universally recognized, however and there remains a great deal of controversy over the amount of damage directly attributable to acid rain. Reduction in forest growth in Sweden (LaBastille 1981), physical damage to trees in West Germany (Pearce 1982b), and the death of sugar maples in Quebec and Vermont (Norton 1985) have all been blamed on the increased acidity of the precipitation in these areas. Although the relationship seems obvious in many cases, the growth, development, and decline of woodland has always reflected the integrated effects of many variables, including site microclimatology, hydrology, land-use change, tree age and species competition. Acid rain has now been added to that list. Even those individuals or groups with greatest concern for the problem admit that it is next to impossible to isolate the impact of any one element from such a combination of variables (Ontario: Ministry of the Environment 1980). Thus it may not be possible to establish definitive proof of the link between acid precipitation and vegetation damage. The body of circumstantial evidence is large, however, and adverse effects have been produced in laboratory experiments. Together, these support the view that the terrestrial environment is under some threat from acid rain.

Assessment of the threat is made difficult by the complexity of the relationships in the terrestrial environment. In areas experiencing acid rain, dry and wet deposition over land is intercepted initially by the vegetation growing there. The effects of this precipitation on the plants may be direct, brought about by the presence of acid particles on the leaves, for example, or indirect, associated with changes in the soil or the biological processes controlling plant growth. Acid precipitation intercepted by trees may promote necrosis of leaf tissue, leaching of leaf nutrients, and chlorophyll degradation (Shriner and Johnston 1985), all of which cause visible damage. Vegetation growing at high altitudes, and therefore enveloped in cloud for long periods, frequently displays such symptoms, since cloud moisture is often more acidic than rain (Hendrey 1985). Ultimately, the acid particles will be washed off the vegetation and into the soil, where they can begin to affect the plants indirectly, but no less seriously.

Once the acid rain enters the soil, its impact will depend very much on the soil type and the underlying bedrock. Soils derived from granite, for example, will already be acidic, and therefore vulnerable to further increases. In contrast, soils developed over limestone, or some other calcium-rich source, will have the ability to neutralize large quantities of additional acid. Natural processes, such as the decay of organic matter or the weathering of minerals, increase the acidity of many soils, and it is often difficult to assess the contribution of acid rain to the total. The situation is further complicated when soils are developed for agriculture. To maintain productivity, it is necessary to make regular applications of fertilizer and lime, which mask acidification (Park 1987).

Acidic water interferes with soil biology and soil chemistry, disturbing nutrient cycles and causing physiological damage to plant-root systems. Increased acidity inhibits the bacterial activity which is instrumental in releasing nutrients from dead or decaying animal and vegetable matter; the ability of nitrifying bacteria to fix atmospheric nitrogen may also be restricted, leading to reduced soil fertility (Ontario: Ministry of the Environment 1980). Changes in soil chemistry, initiated by acid rain, also lead to nutrient depletion. In a normally fertile soil, nutrients such as calcium, potassium, and magnesium are present in the form of positively charged microscopic particles (cations) bonded by way of their electrical charge to clay and humus particles or other soil colloids. They can be removed from there by plants, as required, and are normally replaced by additional cations released into the soil by mineral weathering (Steila 1976). As acidic solutions pass through the soil, hydrogen ions replace the basic nutrient cations, which are then removed in solution with sulphate and nitrate anions (Fernandez 1985). Regular acid-induced leaching of this type, leads to reduced soil fertility, and consequently affects plant growth.

The mobilization of toxic, heavy metals, such as aluminium cadmium, zinc, mercury, lead, copper, and iron, is another feature which accompanies soil acidification (Fernandez 1985). Acid rain liberates the metals from bedrock or soil minerals, and they are carried in solution into ground-water, lakes, and streams, or absorbed by plants. The detrimental effects of heavy metals, such as aluminium, on the aquatic environment, are well-documented. Their impact on the terrestrial environment is less clear, however. The main claims that heavy metals have initiated vegetation damage have come from West Germany, where high levels of aluminium in soils have been blamed, in part, for the forest decline in that area (Fernandez 1985). Elsewhere, there is no clear evidence that tree growth has been impaired by heavy metal mobilization.

The damage attributed to acid rain is both visible and invisible. In some cases, the impact is apparent only after detailed observation and measurement. For example, a survey of annual rings in a mixed spruce, fir, and birch forest exposed to acid rain in Vermont, revealed a progressive reduction in growth rates between 1965 and 1979 (Johnson and Siccama 1983). Between 1965 and 1983, in the same general area, there was also a 25 per cent decline in the above-ground biomass of natural sugar maple forest (Norton 1985). Physical damage to the fine root systems is another element common to trees in areas subject to acid rain (Tomlinson 1985). All of these symptoms are apparent only after careful and systematic survey, but there are other effects which are more directly obvious and which have received most public attention. They can be grouped together under the general term 'tree dieback', which describes the gradual wasting of the tree, inwards from the outermost tips of its branches.

Dieback has been likened to the premature arrival of autumn (Norton 1985). On deciduous trees, the leaves on the outermost branches begin to turn yellow or red in mid-summer; they dry out and eventually fall, well ahead of schedule. These branches will fail to leaf-out in the spring. In succeeding years, the problem will spread from the crown until the entire tree is devoid of foliage, and takes on a skeletal appearance, even in summer (Norton 1985). Coniferous trees react in much the same way. Needles turn yellow, dry up and fall off the branches; new buds fail to open or, if they do, produce stunted and distorted growth (Park 1987). The trees gradually weaken during these changes and become more and more vulnerable to insect attack, disease, and the ravages of weather – all of which contribute to their demise (Norton 1985).

The maple groves of Ontario, Quebec, and Vermont have been suffering progressive dieback since 1980 (Norton 1985), and recent research has shown that the most affected regions are also those that receive the greatest quantities of acid deposition. In Quebec alone, more than 80 per cent of the maple stands show signs of damage (Robitaille 1986). Mortality rates for maples growing in Quebec have increased from 2 per cent per year to 16 per cent (presenting a potential $6 million loss for the province's maple-sugar producers), and regeneration is well below normal (Norton 1985). There is as yet no conclusive proof that acid rain is the cause of dieback, and alternatives such as poor sugar-bush management or disease have been put forward. However, the problem is common to both natural and managed groves, while disease usually follows rather than precedes the onset of dieback. Dieback is now becoming prevalent in beech and white ash stands, but it is the damage to the maple which is causing greatest concern, particularly in Quebec, currently the source of 75 per cent of the world's supply of maple sugar (Norton 1985).

Dieback in coniferous forest has also been identified in eastern North America (Johnson and Siccama 1983), but it is in Europe, specifically West Germany, that it is most extensive. It is estimated that one-third of the trees in that country have already suffered some degree of dieback: 75 per cent of the fir trees and 41 per cent of the spruce trees in the state of Baden-Wurttemberg, which includes the Black Forest, have been damaged (Anon. 1983) and 1,500 hectares of forest have died in Bavaria since the late 1970s (Pearce 1982d). For a nation which prides itself on good forest management, such figures are horrendous. The death of the forests, or *Waldsterben* as it has become known, is progressing at an alarming, and apparently quickening, pace. It is being reported with increasing frequency in Czechoslovakia, East Germany, and Poland (Pearce 1982d), and in Britain the symptoms are becoming clearly recognizable (Park 1987).

Damage to existing trees is only part of the problem. The future of the forests is at stake also. Natural regeneration is no longer taking place in much of central Europe, and even the planting of nursery-raised stock provides no guarantee of success (Tomlinson 1985). Developments such as these raise the spectre of the whole-sale and irreversible loss of forest land. Such is the importance of the forest industry to many of the regions involved, that this would lead to massive economic disruption, and it is therefore not surprising that calls for action on acid rain from these areas are becoming increasingly more strident, perhaps even desperate (Pearce 1982b, Piette 1986).

Acid rain and the built environment

Present concern over acid rain is concentrated mainly on its effect on the natural environment, but acid rain also contributes to deterioration in the built environment. Naturally acid rain has always been involved in the weathering of rocks at the earth's surface. It destroys the integrity of the rock by breaking down the mineral constituents and carrying some of them off in solution. All rocks are affected to some extent, but chalk, limestone, and marble are particularly susceptible to this type of chemical weathering. Inevitably, when these rocks are used as building stone, the weathering will continue. In recent years, however, it has accelerated, in line with the increasing acidity of the atmosphere.

Limestone is a common building stone, because of its abundance, natural beauty, and ease of working. Its main constituents are calcium and magnesium carbonates, which react with the sulphuric acid in acid rain to form the appropriate sulphate. These sulphates are soluble, and are washed out of the stone, gradually destroying the fabric of the building in the process. Further damage occurs when the solutions evaporate. Crystals of calcium and magnesium sulphate begin to form on or beneath the surface of the stone. As they grow, they create sufficient pressure to cause cracking, flaking, and crumbling of the surface, which exposes fresh material to attack by acid rain (Anon. 1984). Limestone and marble suffer most from such processes. Sandstone and granite may become discoloured, but are generally quite resistant to acidity, as is brick. Acid damage to the lime-rich mortar binding the bricks may weaken brick-built structures, however (Park 1987). Structural steel and other metals used in modern buildings may also deteriorate under attack from acid rain (Ontario: Ministry of the Environment 1980).

By attacking the fabric of buildings, acid rain causes physical and economic damage, but it does more than that; it also threatens the world's cultural heritage. Buildings which have survived thousands of years of political and economic change, or the predation of warfare and civil strife, are now crumbling under the attack of acid rain. The treasures of ancient Greece and Rome have probably suffered more damage in the last 50 years than they did in the previous 2,000–3,000 years (Park 1987). On the great cathedrals of Europe – such as those in Cologne, Canterbury, and

Chartres – the craftsmanship of medieval stone masons and carvers may now be damaged beyond repair (Pearce 1982a). Few buildings in the industrialized regions of the world are immune, and damage to the Taj Mahal in India from sulphur pollution may be only the first indication that the problem is spreading to the developing world also (Park 1987).

The chemical processes involved in acid-rain attacks on the built environment are essential'y the same as those involved in the natural environment, but there are some differences in the nature and provenance of the acidity. Damage in urban areas is more often associated with dry deposition than with wet, for example (Park 1987). Those acidic particles which fall out of the atmosphere close to their source of origin, land on buildings, and corrosion begins once moisture is added. Damage is usually attributed to deposition from local sources, such as the smelters or power stations commonly found in urban areas, with little of the long-range transportation associated with acidification in the natural environment. However, in cities with little industrial activity – such as Ottawa, Canada – or in the case of isolated rural structures, most of the damage will be caused by wet deposition originating some distance upwind (LaBastille 1981). Since the various Clean Air Acts introduced in Europe and North America in the 1960s and 1970s had their greatest impact on urban pollution levels, it might be expected that the effects of acid rain on the built environment would be decreasing. There is some indication that this is so, but there appears to be a time lag involved, and it may be some time before the reduction in emissions is reflected in a reduction of acid damage to buildings and other structures (Park 1987).

Acid rain and human health

The infamous London smog of 1952 developed as a result of meteorological conditions which allowed the build-up of pollutants within the urban atmosphere. Smoke – produced by domestic fires, power stations, and coal-burning industries – was the most obvious pollutant, but the most dangerous was sulphuric acid, floating free in aerosol form or attached to the smoke particles (Williamson 1973). Drawn deep into the lungs, the sulphuric acid caused or aggravated breathing problems, and many of the 4,000 deaths attributed to the smog were brought about by the effect of sulphuric acid on the human respiratory system (Bach 1972). Although the Clean Air Acts of the 1960s and 1970s, along with such developments as the tall-stacks policy, reduced the amount of sulphur compounds in urban air, recent studies in Ontario and

Pennsylvania have indicated that elevated atmospheric acidity continues to cause chronic respiratory problems in these areas (Lippmann 1986).

The acid rain which causes respiratory problems is in a dry-gaseous or aerosol form, and mainly of local origin. It is therefore quite different from the far-travelled, wet deposition that has caused major problems in the natural environment. There is, as yet, no evidence that wet deposition is directly damaging to human health, but because of its ability to mobilize heavy metals it may have important indirect effects (Park 1987). For example, heavy metals such as copper, cadmium, zinc, and mercury liberated from soil and bedrock by acid rain may eventually reach the human body via plants and animals in the food chain or through drinking water supplies. The corrosion of storage tanks and distribution pipes by acidified water can also add metals to drinking water. Although quality control in water-treatment plants can deal with such problems (Ontario: Ministry of the Environment 1980), many areas subject to acid rain depend upon wells, springs, and lakes, which provide an untreated water supply. This may expose users to elevated levels of such metals as lead and copper, and although individual doses in all of these situations would be small, regular consumption might allow the metals to accumulate to toxic levels.

Solutions to the problem of acid rain

Although the cause-and-effect relationship between emissions of SO_2 and NO_x and acid-rain damage is not universally accepted, most of the solutions proposed for the problem involve the disruption of that relationship. The basic approach is deceptively simple. In theory, a reduction in the emission rate of acid-forming gases is all that is required to slow down and eventually stop the damage being caused by the acidification of the environment. Translating that concept into reality has proved difficult, however.

The reduction in emissions of SO_2 and NO_x is a long-term solution, based on prevention. It is also a solution in which the environment itself has a major role. As emissions decline, it must adjust until some new level of equilibrium, reflecting the decreased acidity, is attained. There is concern that in some areas the damage has gone too far to be reversed and there is currently some support for this point of view. For example, SO_2 emissions in Britain have declined by nearly 40 per cent since 1970 (Caulfield and Pearce 1984), and, in Ontario, SO_2 loading of the atmosphere declined by 27 per cent in the 5 years between 1972 and 1977 (Ontario:

Ministry of the Environment 1980), yet there is no corresponding reduction in aquatic or terrestrial acidity downwind from these areas. This may indicate that natural recovery is no longer possible, or merely that it is a slow process, taking much longer than the original acidification. It may also indicate that some direct human input will be necessary, either to initiate the recovery process or to speed it up.

One possible input is the addition of lime, which would produce an immediate reduction in acidity, and allow the recovery mechanisms to work more effectively. Lime has been used as a means of sweetening acid soils for many years, and may be the reason that in areas of acid soils agricultural land is less affected by acid rain than the natural environment. In areas where natural regeneration is no longer possible, the restoration of the original chemical balance of the soil by liming and appropriate fertilizer application might allow reforestation to be successful.

The same situation might apply in the aquatic environment. Simply re-stocking an acidified lake with fish cannot be successful unless some buffering agent is added. In 1973, several lakes in the Sudbury area were treated with calcium carbonate and calcium hydroxide in an attempt to reduce acid levels (Scheider *et al.* 1975). Acidity returned to normal and there was an increase in nutrient levels, but, in the lakes closest to Sudbury, copper and nickel remained at concentrations toxic to fish (Ontario: Ministry of the Environment 1980). Similar experiments in Sweden since the mid-1970s have involved the liming of some 1,500 lakes, and have provided encouraging results. Artificial buffering of lakes in this way may be likened to the use of antacid to reduce acid indigestion. The neutralizing effects of the lime may last longer than those of the antacid, but they do wear off in 3–5 years and re-liming is necessary as long as acid loading continues (Ontario: Ministry of the Environment 1980). Thus the treatment of the environment with lime to combat acidity is only a temporary measure, at best. It can be used to initiate recovery, or to control the problem until abatement procedures take effect, but since it deals with only the consequences of acid rain rather than the causes, it can never provide a solution.

Most of the current proposals for dealing with acid rain tackle the problem at its source. They attempt to prevent, or at least reduce, emissions of acid gases into the atmosphere. The only way to stop acid emissions completely is to stop the smelting of metallic ores and the burning of fossil fuels. Modern society could not function without metals, but advocates of alternative energy sources – such as the sun, wind, falling water, and the sea – have long supported a reduction in the use of fossil fuels. These alternative sources can be important locally, but it is unlikely that they will ever have the capacity to replace conventional systems. Nuclear power has also been touted as a replacement, since it can be used to produce electricity without adding gases to the atmosphere. Difficulties associated with the disposal of radioactive wastes remain to be resolved, and events such as the Chernobyl disaster of 1986 do little to inspire public confidence in nuclear power. Thus, although the replacement of fossil-fuel-based energy systems with non-polluting alternatives has the potential to reduce acid rain, it is unlikely to have much effect in the near future.

Since SO_2 makes the greatest contribution to acid rain in North America and Europe, it has received most attention in the development of abatement procedures, whereas emissions of NO_x – which are both lower in volume and more difficult to deal with – have been largely neglected. Similarly the development of control technology has tended to concentrate on systems suitable for conventional power stations, since they are the main sources of acid gases (Kyte 1986a). Sulphur dioxide is formed when coal and oil are burned to release energy, and the technology to control it may be applied before, during, or after combustion. The exact timing will depend upon such factors as the amount of acid reduction required, the type and age of the system, and the cost-effectiveness of the particular process (see Figure 4.6).

One of the simplest approaches to the problem is fuel switching, which involves the replacement of high-sulphur fuels with low-sulphur alternatives. This may mean the use of oil or natural gas rather than coal, but, since most power stations use coal and are not easily converted to handle other fuels, it usually involves the replacement of one type of coal with another or even the blending of low- and high-sulphur coal. Much depends on the availability of the low-sulphur product. In Britain, for example, the supply is limited (Park 1987), but in western Canada and the western United States, abundant supplies of low-sulphur coal are available, with a sulphur content only one-fifth of that which is normal in eastern coal (Cortese 1986). Such a difference suggests that fuel switching has a considerable potential for reducing SO_2 production, yet wholesale substitution is uncommon. The problem is a geographical one. The main reserves of low-sulphur coal in North America are in the west, far removed from the large consumers in the east. Transport costs are therefore high and complete switching becomes economically less attractive than other methods of reducing SO_2 output. Compromise is possible. Rather than switching entirely, Ontario Hydro, the major public electricity producer

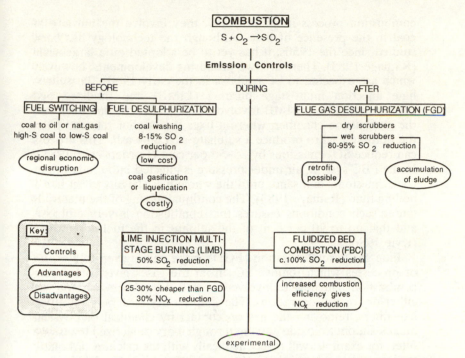

Figure 4.6 Reduction of sulphur dioxide through emission controls

in the province, has a well-established practice of blending low-sulphur, western Canadian coal, with the high-sulphur product from the eastern United States (Ontario: Ministry of the Environment 1980). As a result of this, plus the use of washed coal, the utility's SO_2 output per unit of electricity has been declining with some regularity since the early 1970s (Ontario: Ministry of the Environment 1980).

The amount of SO_2 released during combustion can be reduced if the coal or oil is treated beforehand to remove some of the included sulphur, in a process called fuel desulphurization. The methods can be quite simple and quite cost effective. Crushing and washing the coal, for example, can reduce subsequent SO_2 emissions by 8–15 per cent (Park 1987), which represents a reduction of 1.5–2 million tonnes of SO_2 per year in the eastern United States alone (Cortese 1986).

If it is not possible to reduce sulphur levels significantly prior to combustion, there are techniques which allow reduction during the combustion process itself. Basically, they involve the burning of coal in the presence of lime. Although the technology has been studied since the 1950s, it has yet to be adopted on a large scale (Ramage 1983). There are two promising developments, however, which are expected to be available in the early 1990s. These are: lime injection multi-stage burning (LIMB) and fluidized bed combustion (FBC). LIMB involves the injection of fine lime into the combustion chamber, where it fixes the sulphur released from the burning coal to produce a sulphate-rich lime ash. This process can reduce SO_2 emissions by 35–50 per cent (Burdett *et al.* 1985). In the FBC system, air under pressure is injected into a mixture of coal, limestone, and sand, until the whole mass begins to act like a boiling fluid (Ramage 1983). The continual mixing of the materials under such conditions ensures that combustion is very efficient, and that up to 90 per cent of the sulphur in the fuel is removed (Kyte 1986b).

Flue gas desulphurization (FGD) is the name given to a group of processes which remove SO_2 from the gases given off during combustion. The devices involved are called scrubbers, and may be either dry or wet operations. The simplest dry scrubbers act much like filters, removing the gas on contact by chemical or physical means. Sulphur dioxide passing through a dry pulverized limestone filter, for example, will react chemically with the calcium carbonate to leave the sulphur behind in calcium sulphate (Williamson 1973). Other filters, such as activated charcoal, work by adsorbing the gas on to the filter (Turk and Turk 1988).

Wet scrubbers are more common than the dry variety. The flue gases may be bubbled through an alkaline liquid reagent which neutralizes the SO_2 and produces calcium sulphate in the process (Kyte 1981). A variation on this approach involves the use of a lime slurry through which the flue gases are passed, and in more modern systems the combustion gases are bombarded by jets of lime (LaBastille 1981) or pass through spray systems (Ramage 1983). The system described by LaBastille removes 92 per cent of the sulpher dioxide from the exhaust gases and many scrubbers achieve a reduction of between 80 and 95 per cent (Cortese 1986).

Flue gas desulphurization is one of the most common methods of sulphur dioxide removal, in part because of the high efficiencies possible, but for other reasons also. Scrubbers are technically quite simple, and can be added to existing power plants relatively easily. Retrofitting existing plants in this way is less expensive than building entirely new ones, and in some systems the recovery of sulphuric acid for sale can help to offset the cost.

None of the FGD systems described works well to reduce

emissions of NO_x from power plants. Similar difficulties exist with emission reductions from automobile exhausts. The development of technology that can produce a cooler burning internal combustion engine, or perhaps replace it completely, may be required before emissions of NO_x are reduced significantly (Park 1987).

It is now technically possible to reduce SO_2 emissions to very low levels. However, as is common with many environmental problems, technology is only one of the elements involved. In many cases, economics and politics have retarded the implementation of technical solutions to environmental problems, and that is certainly the case with acid rain.

The economics and politics of acid rain

Government participation in pollution abatement is not a new phenomenon, but, in recent years, particularly following the clean-air legislation of the 1960s and 1970s, the role of government has intensified, and pollution problems have become increasingly a focus for political intervention. Thus, when acid rain emerged as a major environmental problem, it was inevitable that any solution would involve considerable governmental and political activity. Given the magnitude of the problem, it also became clear that it could be solved only at considerable cost, and economic and political factors are now inexorably linked in any consideration of acid rain reduction. Additional complexity is provided by the international nature of the problem.

The costs of acid rain reduction will vary depending upon such factors as the type of abatement equipment required, the reduction of emission levels considered desirable, and the amount of direct rehabilitation of the environment considered necessary. For example one report prepared by the Office of Technology Assessment of the US Congress (Anon. 1984) estimates that a 35 per cent reduction in SO_2 levels in the eastern United States by 1995, would cost between $3 and $6 billion. A 50 per cent reduction in Canada has been estimated to cost $300 million (Israelson 1987). Such costs are not insignificant, and in all likelihood would be imposed, directly or indirectly, on the consumer. Increased costs of this type have to be set against the costs of continuing environmental damage if no abatement is attempted, but the latter often involve less tangible elements, which are difficult to evaluate. The death of a lake from increased acidity, for example, may involve a measurable economic loss through the decline of recreational or commercial fishery (Forster 1985), but there are other items, such as the aesthetic value of the lake, which cannot always be assessed in real monetary terms. Thus, the traditional cost/benefit analysis approach is not always feasible when dealing with the environmental impact and abatement of acid rain.

The first major international initiative to deal with acid rain took place in 1979, when the UN Economic Commission for Europe (UNECE) drafted a Convention on the Long-Range Transportation of Air Pollutants. The Convention was aimed at encouraging a co-ordinated effort to reduce SO_2 emissions in Europe, but the thirty-five signatories also included Canada and the United States (Park 1987). Although not a legally binding document, it did impose a certain moral obligation on the signatories to reduce acid pollution. That was generally insufficient, however, and after additional conferences in Stockholm (1982), Ottawa (1984), and Munich (1984), the UNECE was forced, in 1985, to prepare an additional protocol on sulphur emissions. This was a legally binding document, which required its signatories to reduce transboundary emissions of SO_2 by 30 per cent (of the 1980 levels) before 1993. Fourteen of the original thirty-five participants in the 1979 ECELRTAP Convention refused to sign, among them Britain and the United States (Park 1987). Subsequently, both have become embroiled with neighbouring states, which have signed the protocol, and the resulting confrontation provides excellent examples of the economic and political problems accompanying attempts to reduce acid rain at the international level.

Canada and the United States

Both Canada and the United States are major producers of acid gases (see Figure 4.7). Canada ranks fifth overall in the global SO_2 emissions table (Park 1987), but produces less than one-fifth of the US total. Both countries are also exporters of acid gases, with the United States sending three times as much SO_2 to Canada as Canada sends to the United States (Cortese 1986). It is this discrepancy, and the damage that it causes, which is at the root of the North American acid-rain controversy.

As a member of the so-called '30 per cent club', Canada is obligated to reduce SO_2 emissions by 30 per cent of the 1980 base level before 1993, and has already implemented abatement programmes which will make a 50 per cent reduction possible (Israelson 1987). Such commitments as have been made by the United States have been in the form of financing for increased research. Sulphur-dioxide emissions have been reduced in New England and in the Mid-Atlantic states, but emissions continue to

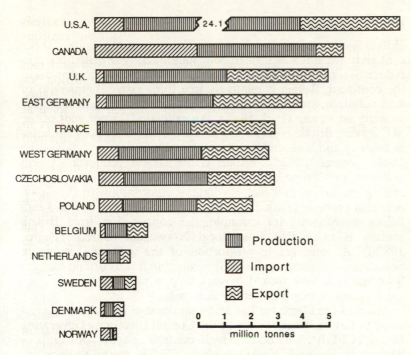

Figure 4.7 The balance of trade in sulphur dioxide in 1980

Source: Calculated from data in Parks (1987); Cortese (1986); Manson (1985); Ontario, Ministry of the Environment (1980)

increase in the mid-west and south-eastern states (Cortese 1986).

The pollution most affecting Canada comes from the mid-west, particularly the Ohio valley, where six states emit more than a million tonnes of SO_2 per year (Israelson 1987). This area, too, is the most resistant to abatement procedures because of their potentially negative impact on a regional economy based on high-sulphur bituminous coal, and its representatives have lobbied strongly and successfully against the institution of emission controls.

Such opposition to acid-rain control has helped to sour relationships between Canada and the United States. Although discussions between the two countries have gone on for over a decade, no substantive agreement on the problem of trans-boundary transportation of acid rain has been reached.

Britain and Scandinavia

Britain was one of the few nations in Europe not to join the '30 per cent club' when it was formed, but subsequently revealed plans to reduce SO_2 emissions by 60 per cent. Situated to the north-west of the continent, Britain is relatively free from external emissions of acid pollution, but it adds some 4.5 million tonnes annually to the westerly air stream (Park 1987). An estimated 25 per cent of the SO_2 leaving Britain is deposited in the North Sea, 4 per cent lands in Norway and Sweden, and 6 per cent continues on to fall in the USSR (Pearce 1982d). Britain is the largest producer of acid pollution in Europe after the USSR; Norway and Sweden are among the lowest and make a limited contribution to their own acid rain problem (Park 1987) (see Figure 4.7). For the acid rain falling on Norway, for example, the contribution from British sources is twice that from local Norwegian sources (Pearce 1982d). A considerable proportion of the acid rain falling in Scandinavia originates in East Germany, but it is to Britain that the Scandinavians have turned to seek help in reducing the environmental damage being caused by acid rain.

Much of the blame for the acid-rain damage in Scandinavia has come to rest on the shoulders of the Central Electricity Generating Board (CEGB), the body which controls electric power production in England. Emissions from CEGB power stations are responsible for more than half the SO_2 produced in Britain (Pearce 1984). Thus, any planned reduction in emissions required the co-operation of the CEGB. The Board, in common with most authorities in Britain, argued that the problem required further study, but that position eventually became untenable. In 1986, for the first time, Britain agreed with Norway that acid rain originating in Britain was causing problems in the Norwegian environment. Subsequently, the British government announced its intentions to cut SO_2 emissions by 14 per cent – commendable, perhaps, but still well below that required for entry into the '30 per cent club' (Park 1987). An agreement reached by the Environment Ministers of the EEC in June 1988 now requires a 60 per cent reduction in emissions by the year 2003, and the CEGB has initiated a series of measures aimed at meeting the new requirements (Kyte 1988).

Summary

Acid rain is a natural product of atmospheric chemical reactions. Inadvertent human interference in the composition of the atmosphere, through the addition of SO_2 and NO_x, has caused it to develop into a major environmental problem. It is mainly confined to the industrialized areas of the northern hemisphere at present but has the potential to become a problem of global proportions. Acid-rain damage is extensive in the aquatic environment and is increasingly recognized in the terrestrial environment. Damage to buildings is common in some areas but the effect of acid rain on human health is less obvious.

Progress towards the large-scale abatement of acid emissions has been slow, and methods for controlling NO_x lag behind those for dealing with SO_2. Emissions of sulphur dioxide are beginning to decline in many areas. Although lakes and forests damaged by acid rain will take some time to recover, action is being taken to improve the situation, and that, in itself, is psychologically important. Finally, much has been written about the necessity to 'clean up' acid rain. Ironically, the acidity is really the end-product of a series of natural cleansing processes by which the atmosphere attempts to maintain some degree of internal chemical balance.

Suggestions for further study

1. Using local weather data, draw a wind-rose to represent the mean annual percentage frequency of the principal wind directions in your area. Map local sources of acid emissions, such as coal-fired power stations or industrial facilities.

   The detailed calculation of effluent dispersal from a pollution source is complex. It depends upon the interaction of a variety of parameters, and the accuracy of the results is limited. However, it is generally accepted that most effluents introduced into the boundary layer fall out soon after release, and individual pollution plumes begin to lose their identity between 5 and 10 km downwind from their source.

   Assuming that wind direction frequency and effluent dispersal are directly related, identify the areas most at risk from acid precipitation within a 10 km radius of the emission sources.

   Acid particles carried into the upper atmosphere will be subject to different wind conditions from those released into the boundary layer. Using the upper-air circulation charts available in most general climatology texts (e.g. Lockwood 1979: 100–3, or Barry and Chorley 1987: 140–1) or, if possible, one obtained from your local weather centre, trace the path taken by locally released acid emissions and identify areas at risk from the long-range transportation of pollutants originating in your area.

2. Examine the buildings in your community and prepare a report on visible evidence of damage caused by acid precipitation. Look for examples of the following: crumbling, peeling, and discoloured stonework; rusted or corroded metalwork; carved inscriptions or coats-of-arms now difficult to read; disfigured statues and other building ornamentation which has suffered damage. Take into consideration such factors as: the age of the building involved; the building material used in the structure; evidence of recent restoration; and the presence of local sources of acid emissions.

Chapter five

Atmospheric turbidity

One of the more obvious indications of atmospheric pollution is the presence of solid or liquid particles, called aerosols, dispersed in the air. These aerosols are responsible for phenomena as diverse as the reduced visibility associated with local-pollution episodes, and the spectacular sunsets which have followed major volcanic eruptions in the past. The attenuation of solar radiation caused by the presence of aerosols provides a measure of atmospheric turbidity, a property which for most purposes can be considered as an indication of the dustiness or dirtiness of the atmosphere.

The concentration and distribution of particulate matter in the atmosphere is closely linked to climatic conditions. Some local or regional climates encourage high aerosol concentrations – as in Los Angeles, for example, with its combination of high atmospheric pressure, light winds, and abundant solar radiation. On a global scale, the mid-latitude westerlies, already implicated in the distribution of acid rain, are responsible for the transportation of aerosols over long distances in the troposphere. The jet streams in the upper atmosphere are also involved in the distribution of aerosols, carrying particles around the world several times before releasing them. Knowledge of such relationships has important practical implications. At the local level, the success of pollution abatement programmes often depends upon an understanding of the impact of climate on aerosol distribution. At a continental, or even hemispheric scale, the relationship between atmospheric circulation patterns and the spread of particulate matter can be used to provide an early warning of potential problems following catastrophic events such as volcanic eruptions or nuclear accidents. In such situations, the atmospheric aerosols are responding to existing climatic conditions. There has been growing concern in recent years that they may do more than that; they may also be capable of initiating climatic change.

97

Aerosol types, production, and distribution

The total global aerosol production is presently estimated to be about 3×10^9 tonnes per annum (Bach 1979). Under normal circumstances, almost all of the total weight of particulate matter is concentrated in the lower 2 km of the atmosphere in a latitudinal zone between 30° N and 60° N (Fennelly 1981). The mean residence time for aerosols in the lower troposphere is between 5 and 9 days, which is sufficiently short that the air can be cleansed in a few days once emissions have stopped (Williamson 1973). The equivalent time in the upper troposphere is about 1 month, and in the stratosphere the residence time increases to 2–3 years (Williamson 1973). As a result, anything added to the upper troposphere or stratosphere will remain in circulation for a longer time, and its potential environmental impact will increase.

Aerosols can be classified in a number of ways, but most classifications include such elements as origin, size, and development – sometimes individually, sometimes in combination (see Figure 5.1). Most of the atmosphere's aerosol content – perhaps as much as 90 per cent (Bach 1979) – is natural in origin, although locally, anthropogenic sources may be dominant, as they are in urban areas, for example (see Figure 5.2). Dust particles created during volcanic activity, or carried into the troposphere during dust storms, are examples of common natural aerosols. Farming practices, quarrying, and a variety of industrial processes contribute dust of anthropogenic origin (Williamson 1973). Natural particles of organic origin – such as those considered responsible for the haze common over such areas as the Blue Mountains of Virginia – join hydrocarbon emissions from human activities to provide an important group of organic aerosols. Total organic emissions in the world have been estimated at as much as 4.4×10^8 tonnes per annum (Fennelly 1981) (1 tonne = .98 tons).

Bolle *et al.* (1986) have grouped aerosols into several mixtures, which include combinations of the different particle types (see Figure 5.1). Three of these are geographically based. In the continental mixture, dust-like particles of natural origin predominate (70 per cent of the total), and most of these are confined to the troposphere. They may be carried over considerable distance, however; dust from the Sahara and Sahel regions of Africa has been carried on the tropical easterlies to the Caribbean (Morales 1986). Logically, volcanically produced aerosols could be included in the continental mixture but, perhaps because of the major contribution of volcanic eruptions to aerosol loading, volcanic ash has been included as a separate item in the classification. In the urban/

AEROSOL CLASSIFICATION

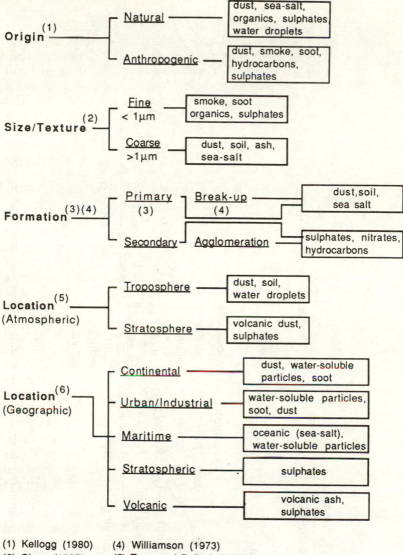

(1) Kellogg (1980)   (4) Williamson (1973)
(2) Shaw (1987)   (5) Toon and Pollack (1981)
(3) Fennelly (1981)   (6) Bolle et al. (1986)

Figure 5.1 A sample of the different approaches used in the classification of atmospheric aerosols

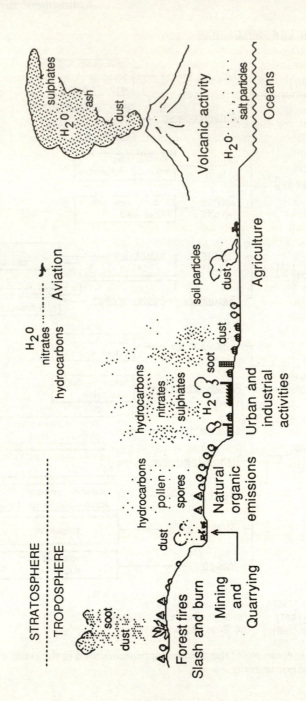

Figure 5.2 Diagrammatic representation of the sources and types of atmospheric aerosols

industrial mix, water-soluble aerosols make up more than 60 per cent of the total. They are of anthropogenic origin, and include a high proportion of sulphates. Soot and water vapour are other important products of urban and industrial activities but mechanically produced dust particles are of relatively minor importance. The effects of individual urban/industrial aerosols may be enhanced by combination with other constituents of the atmosphere. Chemical particles – soot and fine dust, for example – may act as condensation nuclei for the water vapour, and the net result is to increase cloudiness in urban/industrial areas. Aerosols of urban and industrial origin are normally considered to be restricted in their distribution, since they are mainly released into the lower troposphere. However, products of urban combustion found in the Arctic, and the increasing levels of sulphates in the stratosphere, suggest that their effects now extend beyond the local area (Shaw 1980). The third geographically based aerosol mixture recognized by Bolle *et al.* (1986) is maritime in origin. It includes water and sea-spray particles for the most part, with only a small proportion of water-soluble particles (*c.*5 per cent).

Aerosols can range in size from a few molecules to a visible grain of dust, but the distribution across that range is not even. The main mass of particulates is concentrated in two peaks (see Figure 5.3), one between 0.01 micrometer (μm) and 1 μm, centred at 0.1 μm and the other between 1 μm and 100 μm, centred at 10 μm (Shaw 1987). The smaller particles in the first group are called secondary aerosols since they result from chemical and physical processes which take place in the atmosphere. They include aggregates of gaseous molecules, water droplets, and chemical products such as sulphates, hydrocarbons, and nitrates. As much as 64 per cent of total global aerosols are secondary particulates, 8 per cent of them anthropogenic in origin from combustion systems, vehicle emissions, and industrial processes. The other 56 per cent are from natural sources such as volcanoes, the oceans, and a wide range of organic processes (Fennelly 1981). Some estimates suggest that sulphate particles are now the largest group of atmospheric aerosols, accounting for as much as 50 per cent of all secondary particles (Toon and Pollack 1981). Most of the sulphates are of anthropogenic origin, and contribute to the problems of acid rain (see Chapter 4). The larger particles with diameters between 1 and 100 μm are called primary aerosols, and include soil, dust, and solid industrial emissions, usually formed by the physical breakup of material at the earth's surface (Fennelly 1981). There is some evidence that these groupings are a direct result of the processes by which the aerosols are formed. Mechanical processes

101

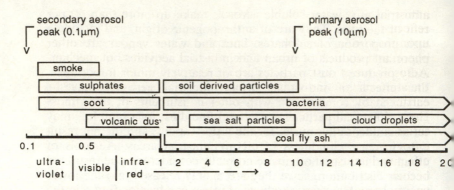

Figure 5.3 A comparison of the size-range of common aerosols with radiation wavelength

are unable to break substances into pieces smaller than 1 μm in diameter, whereas the growth of secondary particles appears to cease as diameters approach 1 μm (Shaw 1987).

Aerosols and radiation

Atmospheric aerosols comprise a very heterogeneous group of particles, and the mix within the group changes with time and place. Following volcanic activity, for example, the proportion of dust particles in the atmosphere may be particularly high; in urban areas, such as Los Angeles, photochemical action on vehicle emissions causes major increases in secondary particulate matter; over the oceans, 95 per cent of the aerosols may consist of coarse sea-salt particles. Such variability makes it difficult to establish the nature of the relationship between atmospheric aerosols and climate. It is clear, however, that the aerosols exert their influence on climate by disrupting the flow of radiation within the earth/ atmosphere system, and there are certain elements which are central to the relationship. The overall concentration of particulate matter in the atmosphere controls the amount of radiation intercepted, while the optical properties associated with the size, shape, and transparency of the aerosols determines whether the radiation is scattered, transmitted, or absorbed (Toon and Pollack 1981).

Several things may happen when radiation strikes an aerosol in the atmosphere. If the particle is optically transparent, the radiant energy passes through unaltered, and no change takes place in the

102

atmospheric energy balance. More commonly, the radiation is reflected, scattered, or absorbed – in proportions which depend upon the size, colour, and concentration of particles in the atmosphere, and upon the nature of the radiation itself (see Figure 5.3). Aerosols which scatter or reflect radiation increase the albedo of the atmosphere and reduce the amount of insolation arriving at the earth's surface. Absorbent aerosols will have the opposite effect. Each process, through its ability to change the path of the radiation through the atmosphere, has the potential to alter the earth's energy budget. The water droplets in clouds, for example, are very effective in reflecting solar energy back into space, before it can become involved in earth/atmosphere processes. Some of the energy scattered by aerosols will also be lost to the system, but a proportion will be scattered forward towards its original destination. Most aerosols – particularly sulphates and fine rock particles – scatter solar radiation very effectively. Only soot particles readily absorb the shorter wave lengths (Toon and Pollack 1981).

The most obvious effects of scattering are found in the visible light sector of the radiation spectrum. Particles in the 0.1 to 1.0 µm size range scatter light in the wavelengths at the blue end of the spectrum, while the red wavelengths continue through. As a result, when the aerosol content of the atmosphere is high, the sky becomes red (Fennelly 1981). This is common in polluted urban areas towards sunset when the path taken by the light through the atmosphere is lengthened, and interception by aerosols is increased. Natural aerosols released during volcanic eruptions produce similar results. The optical effects which followed the eruption of Krakatoa in 1883, for example, included not only magnificent red and yellow sunsets, but also a salmon pink afterglow, and a green colouration when the sun was about 10° above the horizon (Lamb 1970). As well as being aesthetically pleasing, the sequence and development of these colours allowed observers to calculate the size of particles responsible for such optical phenomena (Austin 1983).

In addition to disrupting the flow of incoming solar radiation, the presence of aerosols also has an effect on terrestrial radiation. Being at a lower energy level, the earth's surface radiates energy at the infrared end of the spectrum. Aerosols – such as soot, soil, and dust particles – released into the boundary layer, absorb infrared energy quite readily, particularly if they are larger than 1.0 µm in diameter (Toon and Pollack 1981), and as a result, will tend to raise the temperature of the troposphere. However, since they are almost as warm as the earth's surface, tropospheric aerosols are less efficient at blocking the escape of infrared radiation than colder particles, such as those in the stratosphere (Bolle *et al.* 1986). Thus, longer-wave terrestrial radiation can be absorbed by particles in the stratosphere and re-radiated back towards the lower atmosphere where it has a warming effect. Much depends upon the size of the stratospheric aerosols. If they are smaller in diameter than the wavelengths of the outgoing terrestrial radiation, as is often the case, they tend to encourage scattering and allow less absorption (Lamb 1970). The net radiative effects of particulate matter in the atmosphere are difficult to measure or even estimate. They include a complexity which depends upon the size, shape, and optical properties of the aerosols involved, and upon their distribution in the stratosphere and troposphere.

Any disruption of energy fluxes in the earth/atmosphere system will be reflected ultimately in changing values of such climatological parameters as cloudiness, temperature, or hours of bright sunshine. Although atmospheric aerosols produce changes in the earth's energy budget, it is no easy task to assess their climatological significance. Many attempts at that type of assessment have concentrated on volcanic dust, which for a number of reasons is particularly suitable for such studies. For example, the source of the aerosols can be easily pin-pointed and the volume of material injected into the atmosphere can often be calculated; the dust includes particles from a broad size-range and it is found in both the troposphere and the stratosphere. Recent studies, however, have suggested that the sulphate particles produced during volcanic eruptions have a greater impact on the energy budget than volcanic dust and ash.

Volcanic eruptions and atmospheric turbidity

The large volumes of particulate matter thrown into the atmosphere during periods of volcanic activity are gradually carried away from their sources to be redistributed by the wind and pressure patterns of the atmospheric circulation. Dust ejected during the explosive eruption of Krakatoa, in 1883, encircled the earth in about 2 weeks following the original eruption (Austin 1983), and within 8–12 weeks had spread sufficiently to increase atmospheric turbidity between 35° N and 35° S (Lamb 1970). The diffusion of dust from the Mount Agung eruption in 1963 followed a similar pattern (Mossop 1964) and in both cases the debris eventually spread polewards until it formed a complete veil over the entire earth. The cloud of sulphate particles ejected from El Chichon in 1982 was carried around the earth by the tropical easterlies in less than 20 days, and within a year had blanketed the globe (Rampino and Self 1984).

The build-up of the dust veil and its eventual dispersal will depend upon the amount of material ejected during the eruption and the height to which the dust is projected into the atmosphere. The eruption of Krakatoa released at least 6 km^3 (and perhaps as much as 18 km^3) of volcanic debris into the atmosphere (Lamb 1970). In comparison, Mount St Helens produced only about 2.7 km^3 (Burroughs 1981). Neither of these can match the volume of debris from Tambora, an Indonesian volcano which erupted in 1815, producing an estimated 80 km^3 of ejecta (Findley 1981). More important than the total particulate production, however, is its distribution in the atmosphere. That depends very much on the altitude to which the debris is carried, and whether or not it penetrates beyond the tropopause. The maximum height reached by dust ejected from Krakatoa has been estimated at 50 km and a similar altitude was reached by the dust column from Mount Agung in 1963 (Lamb 1970). A particularly violent eruption at Bezymianny in Kamchatka, in 1956, threw ash and other debris to a height of 45 km (Cronin 1971), but Mount St Helens, despite the explosive nature of its eruption, failed to push dust higher than 20 km, perhaps because the main force of the explosion was directed horizontally rather than vertically (Findley 1981). As a result, it has been estimated that Mount St Helens injected only 5 million tons of debris into the stratosphere compared with 10 million tons for Mount Agung, and as much as 50 million tons for Krakatoa (Burroughs 1981) (1 ton = 1.02 tonnes).

Since the altitude of the tropopause decreases with latitude (see Chapter 2), even relatively minor eruptions may contribute dust to the stratosphere in high latitudes. For example, the dust plume from the Surtsey eruption, off Iceland in 1963, penetrated the tropopause at 10.5 km (Cronin 1971), whereas the products of a comparable eruption in equatorial regions would have remained entirely within the troposphere. Particulate matter which is injected into the stratosphere in high latitudes gradually spreads out from its source, but its distribution remains restricted. Most high-latitude volcanoes are located in a belt close to the Arctic Circle, and there is no evidence of dust from an eruption in this belt reaching the southern hemisphere (Cronin 1971). In contrast, products of eruptions in equatorial areas commonly spread to form a world-wide dust veil (Lamb 1970). As a result of this, it might be expected that when volcanoes are active in both regions, turbidity in the northern hemisphere would be greater than in the southern. Atmospheric turbidity patterns in the period between 1963 and

105

1970, when four volcanic plumes in the Arctic Circle belt and three in equatorial regions penetrated the tropopause, tend to confirm the greater turbidity of the northern stratosphere under such conditions (Cronin 1971).

The dust veil index

Individual volcanic eruptions differ from each other in such properties as: the amount of dust ejected; the geographical extent of its diffusion; and the length of time it remains in the atmosphere. Comparison is possible using these elements, but to simplify the process, and to make it easier to compare the effects of different eruptions on weather and climate, Lamb (1970) developed a rating system which he called a dust veil index (DVI). It was derived using formulae which took into account such parameters as radiation depletion, the estimated lowering of average temperatures, the volume of dust ejected, and the extent and duration of the veil. The final scale of values was adjusted so that the DVI for the 1883 eruption of Krakatoa had a value of 1,000. Other eruptions were then compared to that base. The 1963 eruption of Mount Agung was rated at 800, for example, whereas the DVI for Tambora in 1815 was 3,000 (Lamb 1972).

Individual dust veils may combine to produce a cumulative effect when eruptions are frequent. The 1815 eruption of Tambora, for example, was only the worst of several between 1811 and 1818. The net DVI for that period was therefore 4,400. Similarly, Lamb (1972) estimated that between 1694 and 1698, the world DVI was 3000 to 3500. At times when volcanic activity is infrequent, the DVI is low, as it was between 1956 and 1963 when no eruptions injected debris into the stratosphere (Lockwood 1979).

The DVI provides an indication of the potential disruption of weather and climate by volcanic activity. Dust in the atmosphere reduces the amount of solar radiation reaching the earth's surface, and at high index levels that reduction can be considerable. This is particularly so in higher latitudes, where the sun's rays follow a longer path through the atmosphere and are therefore more likely to be scattered. Major eruptions, producing a DVI in excess of 1,000, have caused reductions in direct-beam solar radiation of between 20 and 30 per cent for several months (Lamb 1972). The effect is diminished to some extent by an increase in diffuse radiation, but the impact on net radiation is negative. Observations in Australia, following the eruption of Mount Agung in 1963, showed a maximum reduction of 24 per cent in direct-beam solar radiation, yet, because of the increase in diffuse radiation, net radiation fell by only 6 per cent (Dyer and Hicks 1965). Similar values were recorded at the Mauna Loa Observatory in Hawaii at that time (Ellis and Pueschel 1971). El Chichon reduced net radiation by 2–3 per cent at ground level (Pollack and Ackerman 1983).

A number of problems with the DVI have been identified since it was first developed, and these have been summarized by Chester (1988). The DVI was based solely on dust and did not include the measurement of sulphates, which are now recognized as being very effective at scattering solar radiation. As a result, the impact of a sulphur-rich eruption such as El Chichon would be under-estimated. At the same time, there is no way of preventing non-volcanic sources of dust from being included in the index. The use of climatic parameters in the calculation of certain index values may introduce the possibility of circular reasoning. For example, falling temperatures are taken as an indication of an increasing DVI, yet a high DVI may also be used to postulate or confirm falling temperatures.

In an attempt to deal with some of these problems, other indices have been proposed, but none has been as widely used as Lamb's DVI. The most recent is the volcanic explosivity index (VEI), developed as a result of research sponsored by the Smithsonian Institute into historic eruptions (Chester 1988). It is based only on volcanological criteria, such as the intensity, dispersive power, and destructive potential of the eruption, as well as the volume of material ejected. It also includes a means of differentiating between instantaneous and sustained eruptions (Newhall and Self 1982). Being derived entirely from volcanological criteria, it eliminates some of the problems of the DVI – such as circular reasoning, for example – but it does not differentiate between sulphates and dust, nor does it include corrections for the latitude or altitude of the volcanoes (Chester 1988). Thus, although the VEI is considered by many to be the best index of explosive volcanism, it is not without its problems when used as a tool in the study of climatic change.

Volcanic activity, weather, and climate

Many of the major volcanic eruptions in historical times have been followed by short-term variations in climate which lasted only as long as the dust veil associated with the eruption persisted. The most celebrated event of this type was the cooling which followed the eruption of Tambora in 1815. It produced in 1816 'the year without a summer', remembered in Europe and North America for its summer snowstorms and unseasonable frosts. Its net effect on world temperature was a reduction of the mean annual value by 0.7 °C, but the impact in mid-latitudes in the northern hemisphere was greater, with a reduction of 1 °C in mean annual temperature, and average summer temperatures in parts of England which were 2–3 °C below normal (Lamb 1970).

Increased volcanic activity may have been a contributing factor in the development of the Little Ice Age – which persisted, with varying intensity, from the mid-fifteenth to the mid-nineteenth century. The eruption of Tambora falls within that time span, and other eruptions have at least a circumstantial relationship with climatic change. A volcanic dust veil may have been responsible for the cool, damp summers and the long, cold winters of the late 1690s in the northern hemisphere – which ruined harvests and led to famine, disease, and an elevated death rate in Iceland, Scotland, and Scandinavia (Parry 1978). Lamb (1970) has suggested that dust veils were important during the Little Ice Age because their cumulative effects promoted an increase in the amount of ice on the polar seas, which in turn disturbed the general atmospheric circulation. He also points out, however, that contemporaneity between increased volcanic activity and climatic deterioration was not complete. Some of the most severe winters in Europe – such as those in 1607–8 and 1739–40 – occurred when the DVI was low, and the period of lowest average winter temperatures did not coincide with the greatest cumulative DVI. Thus, although increased volcanic activity and the associated dust veils can be linked to deterioration between 1430 and 1850. it is likely that volcanic dust was only one of a number of factors which contributed to the development of the Little Ice Age at that time.

Major volcanic episodes in modern times are usually accompanied by prognostications on their impact on weather and climate, although it is not always possible to establish the existence of any cause-and-effect relationship. The Agung eruption produced the second largest DVI this century, but its impact on temperatures was less than expected, perhaps because the dust fell out of the atmosphere quite rapidly (Lamb 1970). It is estimated that it depressed the mean temperature of the northern hemisphere by a few tenths of a degree Celcius for a year or two (Burroughs 1981), but such a value is well within the normal range of annual temperature variation. The spectacular eruption of Mount St Helens in 1980 – enhanced in the popular imagination by intense media coverage – promoted the expectation that it would have a significant effect on climate, and it was blamed for the poor summer of 1980 in Britain. In comparison to other major eruptions in the past, however, Mount St Helens was relatively insignificant in climatological terms. It may have produced a cooling of a few hundredths of a degree Celcius in the northern hemisphere, where its effects would be greatest (Burroughs 1981). The eruption of El Chichon in Mexico in 1982 produced the densest aerosol cloud since Krakatoa, nearly a century earlier. Within a year it had caused global temperatures to decline by at least 0.2 °C and perhaps as much as 0.5 °C (Rampino and Self 1984).

Volcanic activity does have the ability to contribute to changes in weather and climate, although in recent years its impact has been relatively limited. At a time when the human input into global environmental change is being emphasized, it is important not to ignore the contribution of physical processes, such as volcanic activity, which have the ability to augment or diminish the effects of the anthropogenic disruption of the earth/atmosphere system.

The human contribution to atmospheric turbidity

The volume of particulate matter produced by human activities cannot match the quantities emitted naturally (see Figure 5.4). Estimates of the human contribution to total global particulate production vary from as low as 10 per cent (Bach 1979) to more than 15 per cent (Lockwood 1979) with values tending to vary according to the size-fractions included in the estimate. Human activities may provide as much as 22 per cent of the particulate matter finer than 5 μm, for example (Peterson and Junge 1971). It might be expected that the human contribution to atmospheric turbidity would come mainly from industrial activity in the developed nations of the northern hemisphere, but there is some evidence that the burning of tropical grasslands is a much more important source (Bach 1976). Some authorities would also include soil erosion – created by inefficient or destructive agricultural practices – among anthropogenic sources of aerosols (Lockwood 1979).

Various attempts have been made to estimate the impact of human activities on background levels of atmospheric turbidity. One approach to this is to examine turbidity levels in locations such as the high Alps and the Caucasus (renowned for their clean air) or mid-ocean and polar locations (far removed from the common sources of aerosols). These are seen as reference points from which trends can be established. As it stands, the evidence is remarkably contradictory.

There are data which support the view that human activities are enhancing global aerosol levels. A set of observations from Davos in Switzerland suggests an 88 per cent increase in turbidity in the thirty years up to the mid-1960s (McCormick and Ludwig 1967), and dustfall in the Caucasus Mountains showed a rapid increase during a similar time span (Bryson 1968). At Mount St Katherines, over 2,500 m up into the atmosphere in the Sinai, measurements indicate a 10 per cent reduction in direct-beam solar energy up to the mid-1970s, in an area far removed from major industrial sources (Lockwood 1979). The fine particle content of the air above the North Atlantic doubled between 1907 and 1969, and at Mauna Loa in Hawaii, one estimate suggests an increase in turbidity of 30 per cent in the 10 years between 1957 and 1967 (Bryson 1968). That period included the Mount Agung eruption, but the increase was still apparent after the effects of the volcanic emissions had been removed from the calculations (Bryson and Peterson 1968). On the assumption that natural background levels of atmospheric turbidity should remain constant or fluctuate within a relatively narrow range, these rising aerosol levels were ascribed to human activities.

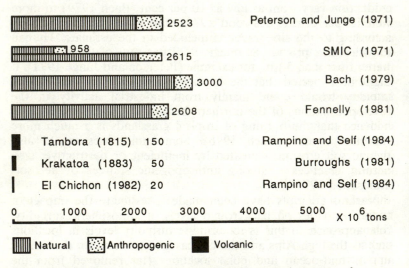

Figure 5.4 A comparison of natural and anthropogenic sources of particulate matter

Note: The volcanic contributions appear small. Their major impact is produced by their ability to make a rapid and intense contribution to the aerosol content of the atmosphere

The development of the Arctic Haze in recent years is generally considered to be an indication of continuing anthropogenic aerosol loading of the atmosphere. The haze is not of local origin. It is created in mid-latitudes as atmospheric pollution, which is subsequently carried polewards to settle over the Arctic. It is most pronounced in winter, for several reasons – including the increased emission of pollutants at that season, the more rapid and efficient poleward transport in winter, and the longer residence-time of haze particles in the highly stable arctic air at that time of year (Shaw 1980). Sulphate particles are the most common constituent of the haze, and their presence at increasingly high concentrations in other remote areas – such as Soviet Georgia – far removed from major pollution sources, is causing concern (Shaw 1987) because of their ability to disrupt the flow of energy in the atmosphere and because of the contribution they make to acid precipitation (see Chapter 4).

Providing a direct contrast to these developments are studies which claim that it is not possible to detect any human imprint in changing atmospheric turbidity. Observations in the Antarctic in the mid-1960s revealed no significant change in aerosol levels in that area between 1950 and 1966 (Fischer 1967). Data from Mauna Loa, where Bryson and Peterson (1968) claimed to find evidence of rising levels of turbidity induced by human activities, were re-interpreted to show that there was no evidence that human activity affected atmospheric turbidity on a global scale (Ellis and Pueschel 1971). The short-term fluctuations revealed in the data were associated with naturally produced aerosols. The contradictory results from Mauna Loa reflect differences in the scientific interpretation of the data from one observatory. Differences between stations are the result of a combination of physical and human factors, and, in reality, are only to be expected. Emissions sites are unevenly distributed. Most are located in mid-latitudes in the northern hemisphere, and there are great gaps – over the oceans, for example – where little human activity takes place. The atmospheric circulation helps to spread the pollutants but, since the bulk of the emissions from anthropogenic sources are confined to the lower troposphere, their residence time in the atmosphere is relatively short. Their eventual distribution is therefore less widespread than volcanic emissions – which are concentrated in the upper troposphere and stratosphere. Aerosols produced in the industrial areas of the north-eastern United States have almost entirely dropped out of the westerly air-stream before it reaches Europe, but aerosols from European and Soviet sources spread over most of North Africa in January, and may drift out over the

Atlantic also, depending upon the strength of the continental high-pressure system (Lockwood 1979). Such patterns may help to explain the rising turbidity levels in the Alps and the Caucasus, and even in the Sinai. In the southern hemisphere, the particulate contribution from industrial activity is negligible, and there is only limited cross-equatorial flow from the north. Aerosols produced during the burning of tropical forests and grasslands will offset the reduction in industrial aerosols (Bach 1976), but the proportion of aerosols of direct anthropogenic origin is usually considered to be much less than in the northern hemisphere. The absence of anthropogenic aerosol sources may explain the lack of change in turbidity over the Antarctic, but the presence of high pressure at the surface and the strong westerlies of the circumpolar vortex aloft may combine to prevent the transport of aerosols into the area. The exact contribution of anthropogenic sources of aerosols to atmospheric turbidity is difficult to determine, and is likely to remain so until the number of measuring sites is increased and their current uneven distribution is improved.

Cooling or warming?

In the mid-1970s, increasing atmospheric turbidity was considered to be one of the mechanisms capable of inducing global cooling (Calder 1974, Ponte 1976). The processes involved seemed plausible and logical, at least in qualitative terms. The introduction of pollutants into the atmosphere, at a rate greater than they could be removed by natural processes, would allow the progressive build-up of aerosols until sufficient quantities had accumulated to cause a rise in global turbidity levels. The net result would be a reduction in insolation values at the earth's surface, as more of the direct-beam solar radiation was scattered or reflected by the particulate matter in the atmosphere. The ability of some of the aerosols to act as condensation nuclei would also tend to increase cloudiness and further reduce the receipt of insolation. From such a scenario it was estimated that an increase of 3–4 per cent in global turbidity levels would be sufficient to reduce the mean temperature of the earth by 0.4 °C, and, according to proponents of anthropogenically produced dust as a factor in climatic change – such as Reid Bryson (1968) – this would be enough to account for the global cooling which took place between 1940 and 1960. Results such as these were based on the observation that global cooling often followed major volcanic eruptions. Particulate matter produced by human activity was considered equivalent to volcanic dust and therefore capable of contributing directly to global cooling. In addition, by elevating background turbidity levels, it allowed smaller volcanic eruptions to be more effective in producing climatic change (Bryson 1968). This comparison of aerosols of human and volcanic origin has been questioned, and it now appears that the cooling ability of aerosols has been exaggerated (Bach 1979, Kellogg 1980, Toon and Pollack 1981).

It has been estimated that natural aerosols in the troposphere probably reduce global surface temperatures by about 1.5 °C (Toon and Pollack 1981), and results obtained by atmospheric-modelling techniques suggest that a doubling of the atmospheric aerosol content would reduce surface temperature by up to 5 °C (Sellers 1973). The assessment of the overall impact has been complicated, however, by indications that the scattering ability of various particles may be less important than their infrared-absorbing properties. The SMIC Report (1971) recognized the ability of particulate matter in the atmosphere to cause warming, but suggested that it was insufficient to compensate for the cooling caused by the attenuation of solar radiation. It is now considered possible that the net effect of elevated atmospheric aerosol levels could be a slight warming, rather than a cooling (Bach 1979).

No realistic value for the impact of anthropogenic aerosols on global temperatures is available. Local and regional changes have occurred, and these indicate that the vertical and horizontal distribution of the aerosols is important. Most particulate matter injected into the atmosphere during human activities does not rise beyond the tropopause. As a result its residence-time is limited, and its impact is confined to an area commonly between 1,000 km and 2,000 km downwind from its source (Kellogg 1980). Most sources of anthropogenic aerosols are on land. There, the addition of particles to the boundary layer tends to reduce the combined albedo of the surface and the lower atmosphere, and the reduced reflection of incoming radiation then promotes warming. The opposite effect is experienced over the oceans, where the combined albedo is increased, producing greater reflectivity and therefore cooling (Bolle *et al.* 1986). Differential changes such as these might in time alter local circulation patterns through their influence on atmospheric stability.

Little anthropogenically produced particulate matter enters the stratosphere at present, but, should that change, the effects would be greater and more prolonged than those produced by the tropospheric aerosols (Bolle *et al.* 1986). Stratospheric aerosols alter the energy budget in two ways. They cause surface cooling by scattering or reflecting incoming solar radiation, but they have the opposite effect on outgoing energy. They absorb terrestrial infrared radiation, which causes a warming of the stratosphere. Together, these intensify the stratospheric temperature inversion, creating greater stability and reducing the vigour of the atmospheric circulation. Experiments with general circulation models (GCM) suggest that an increase in particulate matter in the stratosphere would dampen the Hadley circulation (see Chapter 2), slowing down the easterlies in the tropics and the westerlies in the subtropics (Bolle *et al.* 1986). Records from which aerosol trends can be determined are sparse, and results such as these have been developed mainly through theoretical study rather than by direct observation in the atmosphere. In reality, it is not yet possible to prove that human activities have or have not induced climatic change through the release of aerosols, nor is it possible to make realistic future projections.

Atmospheric turbidity has received less attention from academics and the media in recent years. Perhaps the success of local air-pollution-control measures has helped to reduce the general level of anxiety. In the early 1970s, it seemed possible that anthropogenic aerosols would increase turbidity sufficiently to cause global cooling, and possibly contribute to the development of a new Ice Age (Calder 1974). Concern for cooling has been replaced by concern over global warming, mainly as a consequence of the intensification of the greenhouse effect (see Chapter 7). It has been argued that the presence of particulate matter in the atmosphere has tempered the impact of the greenhouse effect (Bryson and Dittberner 1976), but it now appears possible that under most conditions aerosols actually add to the warming (Bach 1979). Kellogg (1980) has suggested that, on a regional scale, the warming effect of aerosols is more important than the effect of increased carbon dioxide, although, on a global scale, the situation is reversed. He also points out that efforts to control air pollution in industrial areas will ensure that aerosol effects will decline while the impact of the greenhouse effect will continue to grow.

The lack of solid data means that many questions involving the impact of atmospheric turbidity on climate remain inadequately answered. The extent of the human contribution to atmospheric turbidity is still a matter of speculation. Air-pollution monitoring at the local and regional level provides data on changing concentrations of particulate matter over cities and industrial areas, but there is as yet insufficient information to project these results to the global scale. In studying the impact of turbidity on climate, most of the work has dealt with temperature change, but it is also possible that aerosols influence precipitation processes, because of their ability to act as condensation nuclei. The extent and direction of that influence is largely unknown. A general concensus appears to be emerging in the mid-1980s: i.e. that changes in climate brought on by increasing aerosol concentrations have been relatively minor, taking the form of a slight warming rather than the cooling postulated a decade earlier. It is entirely possible, however, that in the event of a significant global-temperature reduction at some time in the future, atmospheric turbidity will be resurrected as a possible cause. The development of new, improved GCMs will help to provide information on the climatic effects of atmospheric aerosols, but it is recognized by the World Meteorological Organization and a number of other international scientific and environmental groups, that direct atmospheric observation and monitoring is essential if the necessary aerosol climatology is to be established (Kellogg 1980, Bolle *et al.* 1986).

Summary

Particulate matter has been a constituent of the atmosphere from the very beginning, and the natural processes which then existed still continue to make the major contribution to atmospheric turbidity. Volcanic activity, dust storms, and a variety of physical and organic processes provide aerosols which are incorporated into the gaseous atmosphere. Human industrial and agricultural activities also help to increase turbidity levels. The aerosols vary in size, shape, and composition – from fine chemical crystals to relatively large, inert soil particles. Once into the atmosphere, they are redistributed by way of wind and pressure patterns, remaining in suspension for periods ranging from several hours to several years, depending upon particle size and altitude attained. The presence of aerosols disrupts the inward and outward flow of energy through the atmosphere. Studies of periods of intense volcanic activity suggest that the net effect of increased atmospheric turbidity is cooling, and some of the coldest years of the Little Ice Age – between 1430 and 1850 – have been correlated with major volcanic eruptions. Aerosols produced by human activities cannot match the volume of material produced naturally. However, in the 1960s and early 1970s, some studies suggested that the cumulative effects of relatively small amounts of anthropogenic aerosols could also cause cooling. Present opinion sees atmospheric turbidity actually producing a slight warming. The greatest problem in the study of the impact of atmospheric turbidity on climate is the scarcity of appropriate data, and that situation can be changed only by the introduction of systematic observation and monitoring, to complement the theoretical analysis – based on atmospheric-modelling techniques – which has been developed in recent years.

Suggestions for further study

1. Put together a list of aerosol sources in your community and its environs. Include natural and anthropogenic sources. Look beyond the more obvious possibilities such as automobile traffic, power stations, or quarries, and consider intermittent sources such as forest fires or wind-eroded soils. Which sources provide most aerosols? What part does local climatology play in the distribution of these aerosols? Do any of the sources make an obvious contribution to global aerosol levels?

2. How can knowledge of the relationship between atmospheric circulation patterns and the spread of particulate matter be used to provide early warning of potential problems following catastrophic events such as volcanic eruptions or nuclear accidents?

3. On an outline map of the world indicate the following: location of the active volcanoes; sources of natural aerosols, e.g. deserts; and sources of anthropogenic aerosols, e.g. major industrial areas.

   Describe the worldwide distribution of these sources.

   Indicate the main, large-scale, wind and pressure belts on the map, and use these to show the most likely spread of aerosols from the following events: a major volcanic eruption in Alaska; dust storms in the Sahel; and a nuclear explosion in central England.

Chapter six

The threat to the ozone layer

One of the most important functions of the atmosphere is to provide the surface of the earth with protection from solar radiation. This may seem contradictory at first sight, since solar radiation provides the energy which allows the entire earth/atmosphere system to function. As with most essentials, however, there are optimum levels beyond which a normally beneficial input becomes harmful. This is particularly so with the radiation at the ultraviolet end of the spectrum. At normal levels, for example, it is an important germicide, and is essential for the synthesis of Vitamin D in humans. At elevated levels it can cause skin cancer, and produce changes in the genetic make-up of organisms. In addition, since ultraviolet radiation is an integral part of the earth's energy budget, changes in ultraviolet levels have the potential to contribute to climatic change.

Under normal circumstances, a layer of ozone gas in the upper atmosphere keeps the ultraviolet rays within manageable limits. This ozone is a relatively minor constituent of the atmosphere. It is diffused through the stratosphere between 10 and 50 km above the surface, reaching its maximum concentration at an altitude of 20 to 25 km. If brought to normal pressure at sea level, all of the existing atmospheric ozone would form a band no more than 3 mm thick (Dotto and Schiff 1978). This small amount of a minor gas, with an ability to filter out a very high proportion of the incoming ultraviolet radiation, is essential for the survival of life on earth. The amount of ozone in the upper atmosphere is not fixed; it may fluctuate by as much as 30 per cent from day to day and by 10 per cent over several years (Hammond and Maugh 1974). Such fluctuations are to be expected in a dynamic system, and are kept under control by built-in checks and balances. By the early 1970s, however, there were indications that the checks and balances were failing to prevent a gradual decline of ozone levels. Inadvertent human interference in the chemistry of the ozone layer was identified as the cause of the decline, and there was growing concern over the potentially disastrous consequences of elevated levels of ultraviolet radiation at the earth's surface. The depletion of the ozone layer became a major environmental controversy by the middle of the decade. Its technological complexity caused dissension in scientific and political arenas, and, with more than a hint of science fiction in its make-up, it garnered lots of popular attention. In common with many environmental concerns of that era, however, interest warned in the late 1970s and early 1980s, only to be revived again with the discovery in 1985 of what has come to be called the Antarctic ozone hole.

The physical chemistry of the ozone layer

The ozone layer owes its existence to the impact of ultraviolet radiation on oxygen molecules in the stratosphere. Oxygen molecules normally consist of two atoms, and in the lower atmosphere they retain that configuration. At the high energy levels associated with ultraviolet radiation in the upper atmosphere, however, these molecules split apart to produce atomic oxygen (see Figure 6.1). Before long, these free atoms combine with the available molecular oxygen to create triatomic oxygen or ozone. That reaction is reversible. The ozone molecule may break down again into its original components – molecular oxygen and atomic oxygen – as a result of further absorption of ultraviolet radiation, or it may combine with atomic oxygen to be reconverted to the molecular form (Crutzen 1974). The total amount of ozone in the stratosphere at any given time represents a balance between the rate at which the gas is being produced and the rate at which it is being destroyed. These rates are directly linked; any fluctuation in the rate of production will be matched by changes in the rate of decay until some degree of equilibrium is attained (Dotto and Schiff 1978). Thus, the ozone layer is in a constant state of flux as the molecular structure of its constituents changes.

The role of ultraviolet radiation and molecular oxygen in the formation of the ozone layer was first explained by Chapman in 1930. Later measurement indicated that the basic theory was valid, but observed levels of ozone were much lower than expected given the limited rate of decay possible through natural processes. Since none of the normal constituents of the atmosphere, such as molecular oxygen, nitrogen, water vapour, or carbon dioxide was considered capable of destroying the ozone, attention was eventually attracted to trace elements in the stratosphere. Initially, it seemed that these were present in insufficient quantity to have the

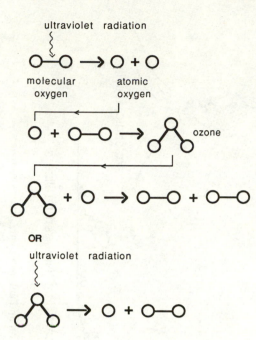

Figure 6.1 Schematic representation of the formation of stratospheric ozone

necessary effect, but the problem was solved with the discovery of catalytic chain reactions in the atmosphere (Dotto and Schiff 1978). A catalyst is a substance which facilitates a chemical reaction, yet remains itself unchanged when the reaction is over. Being unchanged, it can go on to promote the same reaction again and again, as long as the reagents are available, or until the catalyst itself is removed. In this form of chain reaction, a catalyst in the stratosphere may destroy thousands of ozone molecules before it is finally removed. The ozone layer is capable of dealing with the relatively small amounts of naturally occurring catalysts. Recent concern over the thinning of the ozone layer has focused on anthropogenically produced catalysts, which were recognized in the stratosphere in the early 1970s, and which have now accumulated in quantities well beyond the system's ability to cope (see Figure 6.2).

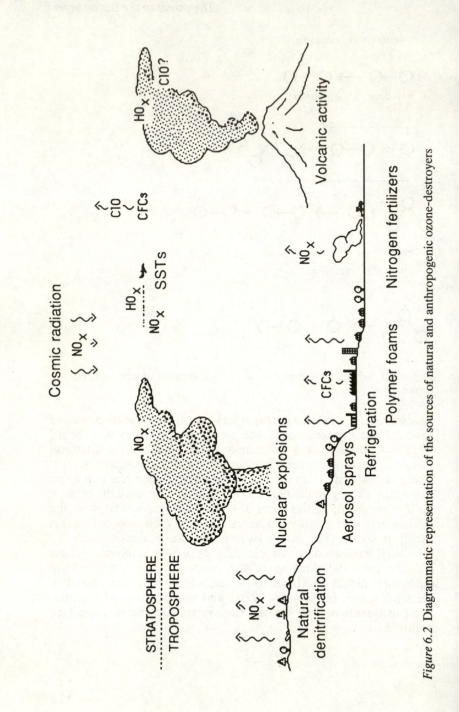

Figure 6.2 Diagrammatic representation of the sources of natural and anthropogenic ozone-destroyers

Naturally occurring, ozone-destroying catalysts

Natural catalysts have probably always been part of the atmospheric system, and many – such as hydrogen, nitrogen, and
chlorine oxides – are similar to those now being added to the
atmosphere by human activities. The main difference is in production and accumulation. The natural catalysts tend to be
produced in smaller quantities and remain in the atmosphere for a
shorter time than their anthropogenic counterparts.

Hydrogen oxides

Hydrogen oxides (HO_x) include atomic hydrogen, the hydroxyl
radical (OH), and the perhydroxyl radical (HO_x) – all derived
from water vapour (H_2O), methane (CH_4), and molecular
hydrogen (H_2), which occur naturally in the stratosphere. They are
referred to collectively as odd hydrogen particles (Crutzen 1972),
and although relatively low in total volume they affect ozone
strongly, particularly above 40 km. It has been estimated that the
HO_x group, through its catalytic properties, is responsible for
about 11 per cent of the natural destruction of ozone in the
stratosphere (Hammond and Maugh 1974). The odd hydrogens
lose their catalytic capabilities when they are converted to water
vapour (see Figure 6.3).

Nitrogen oxides

Nitrogen oxides (NO_x) are very effective destroyers of ozone (see
Figure 6.4). Nitric oxide (NO) is most important, being responsible for 50–70 per cent of the natural destruction of stratospheric
ozone (Hammond and Maugh 1974). It is produced in the stratosphere by the oxidation of nitrous oxide (N_2O), which has been
formed at the earth's surface by the action of denitrifying bacteria
on nitrites and nitrates. It may also be produced in smaller quantities by the action of cosmic rays on atmospheric gases
(Hammond and Maugh 1974). The catalytic chain reaction
created by NO is a long one. Nitric oxide diffuses only slowly into
the lower stratosphere where it is converted into nitric acid, and
eventually falls out of the atmosphere in rain.

Chlorine oxides

The extent to which naturally produced chlorine monoxide (CIO)
contributes to the destruction of the ozone layer is not clear. The

ultraviolet radiation

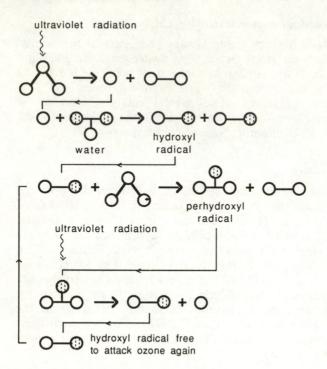

water   hydroxyl radical

perhydroxyl radical

ultraviolet radiation

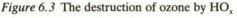

hydroxyl radical free to attack ozone again

Figure 6.3 The destruction of ozone by HO_x

Source: Based on formulae in Crutzen (1972)

most abundant natural chlorine compound is hydrochloric acid (HCl). Although it is present in large quantities in the lower atmosphere, HCl is highly reactive and soluble in water, and Crutzen (1974) considered that it was unlikely to diffuse into the stratosphere in sufficient quantity to have a major effect on the ozone layer. The addition of large amounts of chlorine compounds to the stratosphere during volcanic eruptions was also proposed as a mechanism for the natural destruction of the ozone layer (Stolarski and Cicerone 1974), but observations of the impact of large volcanic eruptions – such as that of Mount Agung (see Chapter 5) – on the ozone layer, do not support that proposition (Crutzen 1974). The impact of naturally occurring ClO on the ozone layer was therefore considered relatively insignificant compared to that of NO_x and HO_x. By 1977, however, measurements showed that the contribution of chlorine to ozone destruction was growing (Dotto and Schiff 1978), but largely from human

122

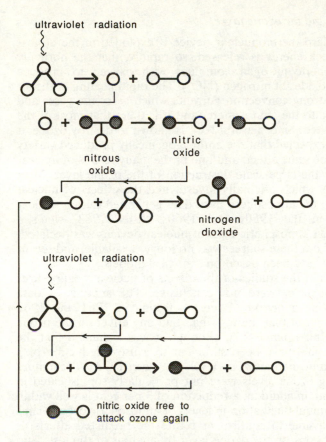

Figure 6.4 The destruction of ozone by NO$_x$

Source: Based on formulae in Crutzen (1972)

rather than natural sources. Anthropogenically produced chlorine now poses a major threat to the ozone layer.

The human impact on the ozone layer

It had become clear by the mid-1970s that human activities had the potential to bring about sufficient degradation of the ozone layer that it might never recover. The threat was seen to come from four main sources – associated with modern technological developments in warfare, aviation, life-style, and agriculture (see Figure 6.2).

123

Nuclear war and the ozone layer

When a modern thermonuclear device is exploded in the atmosphere, so much energy is released, so rapidly, that the normally inert atmospheric nitrogen combines with oxygen to produce quantities of oxides of nitrogen (NO_x). The rapid heating of the air also sets up strong convection currents which carry the gases and other debris into the stratosphere, and it is there that most of the NO_x is deposited. Since natural NO_x is known to destroy ozone, it is only to be expected that the anthropogenically produced variety would have the same effect, and one of the many results of nuclear war might be the large-scale destruction of the ozone layer. Most of the studies which originally investigated the effects of nuclear explosions on the atmosphere used data generated during nuclear bomb tests in the 1950s and 1960s. After 1963, when a moratorium on atmospheric tests of nuclear devices was declared, information from these sources was no longer available, and recent investigations have been based on statistical models.

The results of the studies of the effects of nuclear-weapon tests on the ozone layer were not conclusive. The analysis of data collected during a period of intense testing in 1961 and 1962 produced no proof that the tests had had any effect on the ozone (Foley and Ruderman 1973), although it was estimated that the explosions should have been sufficient to cause a reduction of 3 per cent in stratospheric ozone levels (Crutzen 1974). Techniques for measuring ozone levels were not particularly sophisticated in the 1960s, and, in addition, a reduction of 3 per cent is well within the normal annual fluctuation in levels of atmospheric ozone. Thus it was not possible to confirm or refute the predicted effects of nuclear explosions on the ozone layer by analysis of the test data. Circumstantial evidence did indicate a possible link, however. Since ozone levels are known to fluctuate in phase with sun-spot cycles, it was expected that peak concentrations of ozone in 1941 and 1952 would be followed by a similar peak in 1963, in accordance with the 11-year cycle. That did not happen. Instead, the minimum level reached in 1962 increased only gradually through the remainder of the decade (Crutzen 1974). The missing sun-spot cycle peak was considered to be the result of the nuclear tests, and the gradual increase in ozone in the years following was interpreted as representing the recovery from the effects of the tests, as well as the return to the normal cyclical patterns (Hammond and Maugh 1974).

Although it was not possible to establish conclusive links between nuclear explosions and ozone depletion on the basis of these individual tests, a number of theoretical studies attempted to predict the impact of a full-scale nuclear war on stratospheric ozone. Hampson (1974) estimated that even a relatively minor nuclear conflict, involving the detonation of 50 megatons, would lead to a reduction in global ozone levels of about 20 per cent with a recovery period of several years. He pointed out the importance of thinking beyond the direct military casualties of a nuclear conflict to those who would suffer the consequences of a major thinning of the ozone layer. Since the destruction of the ozone layer would not remain localized, the effects would be felt world-wide, not just among the combatant nations. Subsequent studies by US military authorities at the Pentagon supported Hampson's predictions. They indicated that, following a major nuclear conflict, 50 to 70 per cent of the ozone layer might be destroyed – with the greater depletion taking place in the northern hemisphere where most of the explosions would occur (Dotto and Schiff 1978).

Similar results were obtained by Crutzen (1974) using a photochemical-diffusion model. He calculated that the amount of NO injected into the stratosphere by a 500-megaton conflict would be more than ten times the annual volume provided by natural processes. This was considered sufficient to reduce ozone levels in the northern hemisphere by 50 per cent. Dramatic as these values may appear, they remain approximations – based on data and analyses containing many inadequacies. Interest in the impact of nuclear war on the ozone layer peaked in the mid-1970s and declined thereafter. It emerged again a decade later as part of a larger package dealing with nuclear war and climatology – which emphasized nuclear winter (see Chapter 8).

Supersonic transports and the ozone layer

The planning and development of a new generation of transport aircraft was well under way in North America, Europe, and the USSR by the early 1970s. These were the supersonic transports (or SSTs) designed to fly higher and faster than conventional, subsonic civil airlines, and undoubtedly a major technological achievement. It became clear, however, that they could lead to serious environmental problems if ever produced in large numbers. Initial concerns included elevated noise levels at airports and the effects of the sonic boom produced when the aircraft passed through the sound barrier, but many scientists and environmentalists saw the impact of these high-flying jets on the structure of the ozone layer as many times more serious, and more universal in its effects.

Supersonic transports received a great deal of attention between 1971 and 1974, as a result of Congressional hearings in the United States into the funding of the Boeing SST, and a subsequent climatic impact assessment programme (CIAP) commissioned by the US Department of Transportation to study the effects of SSTs on the ozone layer. The findings in both cases were extremely controversial, and gave rise to a debate which continued for several years, at times highly emotional and acrimonious. It was fuelled further by a series of legal and legislative battles which ended only in 1977, when the US Supreme Court granted permission for the Anglo-French Concorde to land at New York. The proceedings and findings of the congressional hearings and the CIAP, plus the debate that followed, have been summarized and evaluated by Schneider and Mesirow (1976) and Dotto and Schiff (1978). The arguments for and against the SSTs were as much political and economic as they were scientific or environmental. They did reveal, however, a society with the advanced technology necessary to build an SST, yet possessing a remarkably incomplete understanding of the environment into which the aircraft was to be introduced.

Like all aircraft, SSTs produce exhaust gases which include water vapour, carbon dioxide, carbon monoxide, oxides of nitrogen, and some unburned hydrocarbons. These are injected directly into the ozone layer, since SSTs commonly cruise at about 20 km above the surface – just below the zone of maximum stratospheric ozone concentration. Much of the initial concern over the effect of SSTs on ozone centred on the impact of water vapour, which was considered capable of reducing ozone levels through the creation of the hydroxyl radical, a known ozone-destroying catalyst. Later observations, which indicated that a 35 per cent increase in stratospheric water vapour had been accompanied by a 10 per cent increase in ozone, rather than the expected decrease (Crutzen 1972), caused the role of water vapour to be re-evaluated. It was suggested that water vapour helped to preserve the ozone layer through its interaction with other potential catalysts. It converted NO_x to nitric acid, for example, and therefore nullified its ozone-destroying properties (Crutzen 1972, Johnson 1972). By the time this had been confirmed, in 1977, NO_x had already replaced HO_x as the villain in SST operations (Dotto and Schiff 1978).

In 1970, Crutzen drew attention to the role of NO_x in the destruction of ozone through catalytic chain reactions, and in the following year, just as the SST debate was beginning to take off, Johnston (1971) warned that NO_x emitted in the exhaust gases of

500 SSTs could reduce ozone levels by as much as 22–50 per cent. Later predictions by Crutzen (1972) suggested a 3–22 per cent reduction, while Hammond and Maugh reported in 1974 that the net effect of the NO_x emissions from a fleet of 500 SSTs would be a 16 per cent reduction in ozone in the northern hemisphere and an 8 per cent reduction in the southern hemisphere.

When all of this was under consideration in the early 1970s, it was estimated that the world's fleet of SSTs would grow to several hundred aircraft by the end of the century and perhaps to as many as 5,000 by the year 2025 (Dotto and Schiff 1978). The NO_x emissions from such a fleet were considered capable of thinning the ozone layer sufficiently to produce an additional 20,000–60,000 cases of skin cancer in the United States alone (Hammond and Maugh 1974). Other predicted environmental impacts included damage to vegetation and changes in the nature and growth of some species as a result of mutation. The extent to which such threats helped to kill SST development is difficult to estimate. At the time, the environmental arguments seemed strong, but the economic conditions were not really right for development, and that, as much as anything else, probably led to the scrapping of the projected Boeing SST. Development of the Soviet Tupolev-144 and the Anglo-French Concorde went ahead, with the latter being the more successful of the two in terms of production numbers and commercial route development. Less than ten SSTs are currently in operation, and the effects on the ozone layer are generally considered to be negligible.

Chlorofluorocarbons and the ozone layer

If there was some doubt about the impact of SST exhaust emissions on the ozone layer, the effects of some other chemicals seemed less uncertain. Among these, the chlorofluorocarbon (CFC) group has been identified as potentially the most dangerous. The CFCs, sometimes referred to by their trade name, Freon, came to prominence as a result of life-style changes which have occurred since the 1930s. They are used in refrigeration and air-conditioning units but, until recently, their major use was as propellants in aerosol spray cans containing deodorants, hair spray, paint, insect repellant, and a host of other substances. When the energy crisis broke, they were much in demand as foaming agents in the production of polyurethane and polystryrene foam used to improve home insulation. Polymer foams are also included in furniture and car seats, and, with the growth of the convenience food industry, they have been used increasingly in the manufacture

127

of fast-food containers and coffee cups. The gases are released into the atmosphere from leaking refrigeration or air-conditioning systems, or sprayed directly from aerosol cans. They also escape during the manufacture of the polymer foams, and are gradually released as the foams age.

Advantages of CFCs for such purposes include their stability and low toxicity; under normal conditions of temperature and pressure they are inert – that is, they do not combine readily with other chemicals nor are they easily soluble in water. In consequence, they remain in the environment relatively unchanged. Over the years, they have gradually accumulated and diffused into the upper atmosphere where they encounter conditions under which they are no longer inert – conditions which cause them to break down and release by-products which are immensely destructive to the ozone layer (see Figure 6.5).

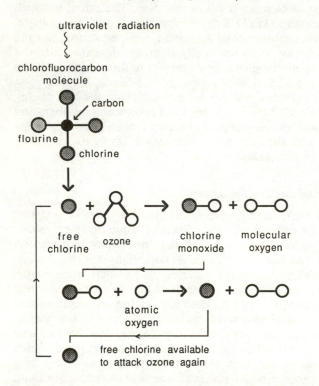

Figure 6.5 The breakdown of a chorofluorocarbon molecule (CFCl₃) and its effect on ozone

In 1974, two scientists working in the United States on the photochemistry of the stratosphere, came to the conclusion that CFCs, however inert they may be at the earth's surface, are highly susceptible to breakdown from the ultraviolet radiation present in the upper atmosphere. Molina and Rowland (1974) recognized that the photochemical degradation of CFCs releases chlorine, which, through catalytic action, has a remarkable capability for destroying ozone. The importance of the chlorine catalytic chain lies in its efficiency; it is six times more efficient catalytically than the NO cycle, for example. The chain is broken only when the Cl or ClO gains a hydrogen atom from the odd hydrogen group, or from a hydrocarbon such as methane (CH_4), and is converted into HCl, which diffuses into the lower atmosphere, eventually to be washed out by rain (Hammond and Maugh 1974). Similar conclusions were reached independently at about the same time by other researchers (Crutzen 1974, Cicerone *et al.* 1974, Wofsy *et al.* 1975), and, with the knowledge that the use of CFCs had been growing since the late 1950s, the stage seemed set for an increasingly rapid thinning of the earth's ozone shield, followed by a rise in the level of ultraviolet radiation reaching the earth's surface.

The world production of CFCs reached 700,000 tons in 1973 (Crutzen 1974), after growing at an average rate of about 9 per cent in the 1960s (Molina and Rowland 1974). The effects of such increases in production were exacerbated by the stability of the product, which allows it to remain in the atmosphere for periods of 40–150 years, and measurements in the troposphere in the early 1970s indicated that almost all of the CFCs produced in the previous two decades were still there (Molina and Rowland 1974). This persistence means that, even after a complete ban on the production of CFCs, the effects on the ozone layer may continue to be felt for a minimum of 20 to 30 years (Crutzen 1974) and, under certain circumstances, for as long as 200 years (Wofsy *et al.* 1975).

The predicted effects of all of this on the ozone layer varied. Molina and Rowland (1974) estimated that the destruction of ozone by chlorine was already equivalent to that produced by naturally occurring catalysts. Crutzen (1974) predicted that a doubling of CFC production would cause a corresponding 10 per cent reduction in ozone levels, whereas Wofsy *et al.* (1975) estimated that, with a growth rate of 10 per cent per year, CFCs could bring about a 20 per cent reduction by the end of the century. All indicated the preliminary nature of their estimates and the inadequacy of the existing knowledge of the photochemistry of the stratosphere, but such cautions were ignored as the topic took on a momentum of its own, and the dire predictions made following SST studies were repeated. The spectre of thousands of cases of skin cancer linked to a seemingly innocuous product like hairspray or deodorant was sufficiently startling that it excited the media and, through them, the general public. Although CFCs were being employed as refrigerants and used in the production of insulation, the problem was usually presented as one in which the convenience of aerosol-spray products was being bought at the expense of the global environment. In 1975 there was some justification for this, since at that time 72 per cent of CFCs were used as propellants in aerosol spray cans (Webster 1988), and the campaign against that product grew rapidly.

The multi-million-dollar aerosol industry, led by major CFC producers such as DuPont, reacted strongly. Through advertising and participation in US government hearings, they emphasized the speculative nature of the Molina–Rowland hypothesis, and the lack of hard scientific facts to support it. The level of concern was high, however, and the anti-aerosol forces met with considerable success. Eventually manufacturers were forced to replace CFCs with less hazardous propellants (Dotto and Schiff 1978). A partial ban on CFCs, covering their use in hair and deodorant sprays was introduced in the United States in 1978 and in Canada in 1980. CFC aerosol spray use remained high in Europe where there was no ban. In 1989, however, the European Community agreed to eliminate the production and use of CFCs by the end of the century.

The CFC controversy had ceased to make headlines by the late 1970s, and the level of public concern had fallen away. Monitoring of the ozone layer showed little change. Ozone levels were not increasing despite the ban on aerosol sprays, which was only to be expected given the slow rate of decay of the existing CFCs, but the situation did not seem to be worsening. Quite unexpectedly, in 1985, scientists working in the Antarctic announced that they had discovered a 'hole' in the ozone layer, and all of the fears suddenly returned.

The Antarctic ozone hole

Total ozone levels have been measured at the Halley Bay base of the British Antarctic Survey for 30 years, beginning in the late 1950s. Seasonal fluctuations were observed for most of that time, and included a thinning of the ozone above the Antarctic during the southern spring, which was considered part of the normal variability of the atmosphere (Schoeberl and Krueger 1986). This

130

regular minimum in the total ozone level began to intensify in the early 1980s, however (see Figure 6.6). Farman *et al.* (1985) reported that it commonly became evident in late August, and got progressively worse until, by mid-October, as much as 40 per cent of the ozone layer above the Antarctic had been destroyed. Usually the hole would fill by November, but during the 1980s it began to persist into December. The intensity of the thinning and its geographical extent were originally established by ground based measurements, and later confirmed by remote sensing from the Nimbus-7 polar-orbiting satellite (Stolarski *et al.* 1986).

As was only to be expected after the aerosol-spray-can experience in the 1970s, the immediate response was to implicate CFCs. Measurements of CFCs at the South Pole indicating a continuous increase of 5 per cent per year tended to support this, although the original fluctuation had been present before CFCs were released into the atmosphere in any quantity (Schoeberl and Krueger 1986). Other researchers have suggested that gases such as NO_x (Callis and Natarajan 1986) and the oxides of chlorine (ClO_x) and

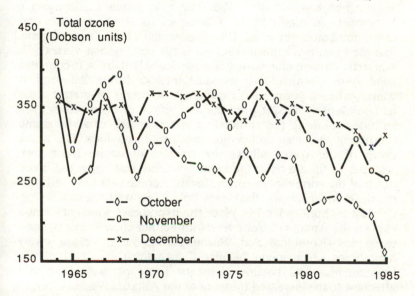

Figure 6.6 Changing ozone levels at the South Pole (1964–85) for the months of October, November, and December

Source: Compiled from data in Komhyr *et al.* (1986)
Note: Dobson units (DU) are used to represent the thickness of the ozone layer at standard (sea-level) temperature and pressure (1 DU is equivalent to 0.01 mm)

bromine (BrO_x) (Crutzen and Arnold 1986) are the culprits. As the investigation of the Antarctic ozone layer intensifies, it is becoming clear that the chemistry of the polar stratosphere is particularly complex. Although the stratosphere is very dry, it becomes saturated at the very low temperatures reached during the winter months, and clouds form. These include nitric and hydrochloric acid particles, which through a complex series of reactions – involving denitrification and dehydration – cause chlorine to be released. In the form of ClO it then attacks the ozone (Shine 1988). The evaporation of the polar stratospheric clouds in the spring, as temperatures rise, brings an end to the reactions, and allows the recovery of the ozone layer.

In contrast to these approaches which emphasize the chemistry of the stratosphere, there are the so-called dynamic hypotheses, which seek to explain the variations in ozone levels in terms of circulation patterns in the atmosphere. Certain observations do support this. For example, at the time the ozone level in the hole is at its lowest, a ring of higher concentration develops around the hole at between 40° and 50° S. The hole begins to fill again in November, seemingly at the expense of the zone of higher concentration (Stolarski *et al.* 1986). Bowman (1986) has suggested that the main mechanism involved is the circumpolar vortex. The Antarctic circumpolar vortex is a particularly tight, self-contained wind system, which is at its most intense during the southern winter, when it permits little exchange of energy or matter across its boundaries. Thus, it prevents any inflow of ozone from lower latitudes, allowing the cumulative effects of the catalytic ozone-destroying chemicals to become much more obvious. The breakdown of the vortex allows the transfer of ozone from lower latitudes to fill the hole. Observations show that when the breakdown of the vortex is late, ozone levels become very low, and when the breakdown is early, the ozone hole is less well marked.

Most authorities tend to place the hypotheses which attempt to explain the Antarctic ozone levels into either chemical or dynamic categories (Rosenthal and Wilson 1987), but there are several hypotheses which might be placed in a third group, combining both chemical and dynamic elements. The paper which first drew attention to the increased thinning of the Antarctic ozone layer, for example, might be classed in the chemical group since it attributes the decline to CFCs (Farman *et al.* 1985). It does, however, have a dynamic element in its consideration of the Antarctic circumpolar vortex, which appears to be a necessary prerequisite for the ozone depletion. The circumpolar vortex in the northern hemisphere is much less intense than its southern counterpart – which may explain, in part at least, the absence of a comparable hole in the ozone over the Arctic (Farman *et al.* 1985). A smaller, more mobile hole has been identified in the ozone above the Arctic, however, and further investigation is in progress (Shine 1988).

Another hypothesis which combines dynamic and chemical elements is that proposed by Callis and Natarajan (1986). They suggest that the depletion of the ozone in the Antarctic is a natural phenomenon, caused by elevated levels of NO_x in the atmosphere during periods of increased solar activity. Sun-spot activity did peak at a particularly high level in 1979, and continued into the early 1980s, but recent measurements by Dr Susan Solomon, released at a special meeting of the Royal Meteorological Society in 1988, indicate that NO_x levels in the lower stratosphere in the Antarctic are too low to have the necessary catalytic effect (Shine 1988).

There is as yet no one theory which can explain the creation of the Antarctic ozone hole adequately. The impact of CFCs is considered the most likely cause by many, however. There can be little doubt that the link between the ozone hole and the CFCs, tenuous as it may seem to some scientists, helped to revive environmental concerns and contributed to the speed with which the world's major industrial nations agreed in Montreal in 1987, to take steps to protect the ozone layer.

Agricultural fertilizers, nitrous oxide, and the ozone layer

When concern for the ozone layer was at its height, compounds other than CFCs – produced in increasing quantities by human activities – were identified as potentially harmful. These included nitrous oxide, carbon tetrachloride, and chloroform, with the first of these probably most likely to cause problems. Nitrous oxide (N_2O) is produced naturally in the environment by denitrifying bacteria which cause it to be released from the nitrites and nitrates in the soil. It is an inert gas, not easily removed from the troposphere. Over time, it gradually diffuses into the stratosphere where the higher energy levels cause it to be oxidized into nitric oxide (NO), leading to the destruction of ozone molecules. This process was part of the earth/atmosphere system before human beings came on the scene, and, with no outside interference, natural ozone levels adjust to the output of N_2O from the soil and the oceans.

One of society's greatest successes had been the propogation of the human species, as a glance at world population growth will show. This has occurred for a number of reasons, but would not have been possible without a growing ability to supply more and more food as population numbers grew. By the late 1940s and 1950s, this ability was being challenged as population began to outstrip food supply. In an attempt to deal with the problem, new agricultural techniques were introduced into Third World countries, where the need was greatest. A central element in the process was the increased use of nitrogen fertilizers along with genetically improved grains, which together produced the necessary increase in agricultural productivity. Since that time, continued population growth has been paralleled by the growth in the use of nitrogen based fertilizers (Dotto and Schiff 1978).

The nitrogen in the fertilizer used by the plants eventually works its way through the nitrogen cycle, and is released into the air as N_2O to initiate the sequence which ultimately ends in the destruction of ozone. Thus, in theory, the pursuit of greater agricultural productivity through the increased application of nitrogen fertilizers is a threat to the ozone layer. There is, however, no proof that increased fertilizer use has, or ever will, damage the ozone layer, through the production of N_2O. If proof does emerge, it will create a situation not uncommon in humankind's relationship with the environment, in which a development designed to combat one problem leads to others, unforeseen and perhaps undiscovered until major damage is done. As Schneider and Mesirow (1976) point out, the dilemma lies in the fact that nitrogen fertilizers are absolutely essential to feed a growing world population, and improve its quality of life, yet success might lead to a thinning of the ozone layer with consequent climatic change and biological damage from increased ultraviolet radiation. If monitoring does reveal that N_2O emissions are increasing, some extremely difficult judgements will have to be made: judgements which could have a far greater impact than the grounding of SSTs or the banning of CFCs from use in aerosol spray cans.

The biological and climatological effects of changing ozone levels

Declining concentrations of stratospheric ozone allow more ultraviolet radiation to reach the earth's surface. Even after a decade and a half of research, the impact of that increase is still very much a matter of speculation, but most of the effects which have been identified can be classified as either biological or climatological.

Biological effects

In moderate amounts ultraviolet radiation has beneficial effects for life on earth. It is a powerful germicide, for example, and triggers the production of Vitamin D in the skin. Vitamin D allows the body to fix the calcium necessary for proper bone development; lack of it may cause rickets, particularly in growing children. High intensities of ultraviolet radiation, however, are harmful to all forms of life.

Life forms on earth have evolved in such a way that they can cope with existing levels of ultraviolet radiation. They are also quite capable of surviving the increases in radiation caused by short-term fluctuations in ozone levels. Most organisms would be unable to cope with the cumulative effects of progressive thinning of the ozone layer, however, and the biological consequences would be far-reaching.

Intense ultraviolet radiation is capable of damaging the basic foundations of life, such as the DNA molecule and various proteins (Crutzen 1974). It also inhibits photosyntheis. Growth rates in plants such as tomatoes, lettuce, and peas are reduced, and experimental exposure of some plants to increased ultraviolet radiation has produced an increased incidence of mutation (Hammond and Maugh 1974). Insects, which can see in the ultaviolet sector of the spectrum, would have their activities disrupted by increased levels of ultraviolet radiation (Crutzen 1974).

Most of the concern for the biological effects of declining ozone levels has been focused on the impact of increased ultraviolet radiation on the human species. The potential effects include the increased incidence of sunburn, premature ageing of the skin among white populations, and greater frequency of allergic reactions caused by the effects of ultraviolet light on chemicals in contact with the skin (Hammond and Maugh 1974). These are relatively minor, however, in comparison to the more serious problems of skin cancer and radiation blindness, both of which would become more frequent with higher levels of ultraviolet radiation. Skin cancer had a prominent role in the SST and CFC debates of the 1970s (Dotto and Schiff 1978), and it continues to evoke a high level of concern. The number of additional skin cancers to be expected as the ozone layer thins is still a matter of debate. Hammond and Maugh (1974) have suggested that a 5 per cent reduction in ozone levels would cause an additional 20,000 to 60,000 skin cancers in the United States alone. The US National Academy of Sciences has also estimated that a 1 per cent decline in ozone would cause 10,000 more cases of skin cancer per year

(Lemonick 1987). Many skin cancers are non-malignant and curable, but only after painful and sometimes disfiguring treatment. A relatively small proportion are malignant and usually fatal (Dotto and Schiff 1978). The Environmental Protection Agency (EPA) forecasts that 39 million more people than normal would contract skin cancer within the next century, leading to more than 800,000 additional deaths (Chase 1988).

The attention paid to skin cancer has caused other effects to be overshadowed. Radiation blindness and cataracts were early identified as potential problems (Dotto and Schiff 1978). More recently, damage to the human immune system has been postulated, and there is some evidence that ultraviolet light may be capable of activating the AIDS virus (Valerie *et al.* 1988).

Concentration on the direct effects of ozone depletion on people is not surprising or unexpected. Much more research into other biological effects is required, however. Human beings are an integral part of the earth/atmosphere system and, as Dotto and Schiff (1978) suggest, humankind may experience the consequences of ozone depletion through its effects on plants, animals, and climate. The impact would be less direct, but perhaps no less deadly.

Climatological effects

The climatological importance of the ozone layer lies in its contribution to the earth's energy budget (see Figure 6.7). It has a direct influence on the temperature of the stratosphere through its ability to absorb incoming radiation. Indirectly, this also has an impact on the troposphere. The absorption of short-wave radiation in the stratosphere reduces the amount reaching the lower atmosphere, but the effect of this is limited to some extent by the emission of part of the absorbed short-wave energy into the troposphere as infrared radiation.

Natural variations in ozone levels alter the amounts of energy absorbed and emitted, but these changes are an integral part of the earth/atmosphere system, and do little to alter its overall balance. In contrast, chemically induced ozone depletion could lead to progressive disruption of the energy balance, and ultimately cause climatic change. The total impact would depend upon a number of variables, including the amount by which the ozone concentration is reduced, and the altitude at which the greatest depletion occurred (Schneider and Mesirow 1976).

A net decrease in the amount of stratospheric ozone would reduce the amount of ultraviolet absorbed in the upper atmosphere,

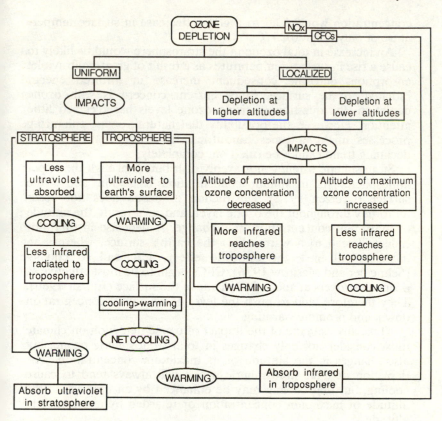

Figure 6.7 Schematic representation of the radiation and temperature changes accompanying the depletion of stratosphere ozone

producing cooling in the stratosphere. In theory, the radiation no longer absorbed would continue on to the earth's surface, causing the temperature there to rise. This simple response to declining ozone concentration is complicated by the effects of stratospheric cooling on the system. The lower temperature of the stratosphere causes less infrared radiation to be emitted to the troposphere, and the temperature of the lower atmosphere falls also. Since the cooling effect of the reduction in infrared energy is greater than the warming caused by the extra short-wave radiation, the net result is a cooling at the earth's surface. The magnitude of the cooling is difficult to assess, but it is likely to be small. Enhalt (1980) has suggested that a 20 per cent reduction in ozone

137

concentration would lead to a global decrease in surface temperature of only about 0.25 °C.

An increase in total ozone in the stratosphere would be likely to cause a rise in surface temperatures as a result of greater ultraviolet absorption, and the consequent increase in infrared energy radiated to the surface. Since current concern is with ozone depletion, the question of rising ozone levels has received little attention. However, the possibility that natural, ozone-enhancing processes might at times be sufficiently strong to reverse the declining trend cannot be ruled out completely.

Stratospheric ozone is not evenly distributed through the upper atmosphere. Its maximum concentration is 25 km above the surface (Crutzen 1972). Destruction of ozone does not occur uniformly throughout the ozone layer, and, as a result, the altitude of maximum concentration may change. A decrease in that altitude will lead to a warming of the earth's surface, whereas an increase will have the opposite effect, and lead to cooling (Schneider and Mesirow 1976). CFCs begin to be most effective as ozone destroyers at about 25 km above the surface (Enhalt 1980). They therefore tend to push the level of maximum concentration down, and promote warming.

Thus, any estimate of the impact of ozone depletion on climate must consider not only changes in total stratospheric ozone, but also changes in the altitude of its maximum concentration. The depletion of total stratospheric ozone will always tend to cause cooling, but that cooling may be enhanced by an increase in the altitude of maximum concentration or retarded by a decrease in altitude.

Just as there are variations in the vertical distribution of ozone, there are also variations in its horizontal distribution. The latter are seasonal and associated with changing wind and pressure systems in the lower stratosphere (Crutzen 1974). Most ozone is manufactured above the tropics, and is transported polewards from there. Increased levels of ozone have been identified regularly at middle and high latitudes in late winter and spring in the northern hemisphere (Crutzen 1972), and the redistribution of ozone by upper atmospheric winds has been implicated by a number of authors in the development of the Antarctic ozone hole (Shine 1988). Such changes have local and short-term effects which might reinforce or weaken the global impact of ozone depletion.

Further complications are introduced by the ability of several of the chemicals which destroy ozone to interfere directly with the energy flow in the atmosphere. Ramanathan (1975) has shown that ozone-destroying CFCs absorb infrared radiation, and the resulting temperature increase might be sufficient to negate the cooling caused by ozone depletion. Similarly, oxides of nitrogen absorb solar radiation so effectively that they are able to reduce the cooling caused by their destruction of ozone by about half (Ramanathan *et al.* 1976).

Changes in the earth's energy budget initiated by declining stratospheric ozone levels are integrated with changes produced by other elements such as atmospheric turbidity and the greenhouse effect. The specific effects of ozone depletion are therefore difficult to identify, and the contribution of ozone depletion to climatic change difficult to assess.

The Montreal Protocol

In September 1987, thirty-one countries, meeting under the auspices of the United Nations Environmental Programme in Montreal, signed an agreement to protect the earth's ozone layer. The Montreal Protocol was the culmination of a series of events which had been initiated two years earlier at the Vienna Convention for the Protection of the Ozone Layer. Twenty nations signed the Vienna Convention in September 1985, promising international co-operation in research, monitoring, and the exchange of information on the problem. In the two-year period between the meetings much time and effort went into formulating plans to control the problem, with the countries of the European Economic Community (EEC) favouring a relatively gradual approach compared to the more drastic suggestions of the North Americans (Tucker 1987). The Environmental Protection Agency (EPA) in the United States, against a background of an estimated 39 million additional cases of skin cancer in the next century, suggested a 95 per cent reduction in CFC production within a period of 6–8 years (Chase 1988). When the Montreal Protocol was signed, participants agreed to a 50 per cent production cut by the end of the century, although that figure is deceptive, since Third World countries will be allowed to increase their use of CFCs for a decade to allow technological improvements in such areas as refrigeration. The net result will be only a 35 per cent reduction in total CFC production by the end of the century, based on 1986 totals (Lemonick 1987). This may be a historic agreement, but certain experts in the field, such as Sherwood Rowland, feel that it is not enough, and that the original 95 per cent proposed by the EPA was absolutely necessary (Lemonick 1987).

The signatories to the Montreal Protocol met again in Helsinki in May 1989. At that time, along with participants from some 50 additional countries, they agreed to the elimination of CFC production and use by the year 2000.

In contrast to their attitudes in the 1970s, the CFC manufacturers have responded positively to the call for a reduction in the manufacture of the chemical. The DuPont Company, for example, pledged to reduce its output by 95 per cent by the year 2000 (the original EPA suggestion), although the initial search for appropriate substitutes – which includes testing for health and environmental effects – may take up to 5 years (Anon. 1988). The market for substitutes is large, particularly in Europe, where CFCs were not banned in the 1970s. Although the concern for the supply of energy is no longer at crisis levels, the demand for residential and industrial building insulation remains high, and will undoubtedly rise if energy supplies are again threatened. Since the manufacture of insulating materials accounts for 28 per cent of worldwide CFC production, the search for alternatives is receiving urgent attention. Results have been mixed. DuPont, for example, has developed a hydrochlorofluorocarbon (HCFC) as a potential replacement for polystyrene sheet. It is 95 per cent less damaging to ozone than normal CFC because it is less stable and tends to break down in the troposphere before it can diffuse into the ozone layer. It has been used successfully in food packaging, but unfortunately it is not suitable for home insulation since it loses its insulating ability as the HCFCs break down (Webster 1988). An isocyanate-based insulation foam, in which carbon dioxide is the foaming agent, has produced promising results. It cannot be produced in sheet form, however, and, as a result, its use remains limited to situations where spray-on or foam-in application is possible. Thus, the CFC manufacturing companies are committed to the search for alternatives, but it may be some time before their good intentions are translated into a suitable product.

Summary

There is an ever-increasing amount of evidence that the earth's ozone layer is being depleted, allowing a higher proportion of ultraviolet radiation to reach the earth's surface. If allowed to continue, this would cause a serious increase in skin cancer cases, produce more eye disease, and change the genetic make-up of terrestrial organisms. In addition, since ultraviolet radiation is an integral part of the earth's energy budget, any increase in its penetration to the lower atmosphere could lead to climatic change. The effects of a depleted ozone layer are widely accepted, but the extent of the depletion and its cause remain matters of controversy. This reflects society's inadequate knowledge of the workings of the earth/atmosphere system in general, and the photochemistry of the stratosphere in particular. It may take many years of observation and research before this situation is altered, but in the meantime a general concensus among scientists and politicians, that CFCs are the main culprits in the destruction of the earth's ozone, has led to the proscription of that group of chemicals. Eventually, CFCs will no longer be produced, but their effects will linger on until those presently in the atmosphere are gone.

In dealing with the global aspects of air pollution involving such elements as acid precipitation, atmospheric turbidity, and the threat to the ozone layer, it is quite clear that although society now has the ability to cause all of these problems, and may even possess the technology to slow down and reverse them, its understanding of their overall impact on the earth/atmosphere system lags behind. Until that can be changed, the effects of human activities on the system will often go unrecognized, response to problems will of necessity be reactive, and the damage done before the problem is identified and analysed may be irreversible.

Suggestions for further study

1. It is a scientific axiom that energy can neither be created nor destroyed. Ultraviolet radiation is a form of energy, prevented from reaching the earth's surface by the ozone layer. If the energy in the ultraviolet radiation is not destroyed, what happens to it?

2. In the upper atmosphere the destruction of ozone is considered harmful, yet at the earth's surface it is the production of ozone that is considered harmful. Why does this apparent contradiction exist? Why does the ozone produced in the lower atmosphere not replace that destroyed in the upper atmosphere?

3. One scientist has claimed that if all CFCs were burned completely and immediately, probably more people would die from food poisoning as a consequence of inadequate refrigeration than would die from depleting ozone (Chase 1988). Using the relationship between CFCs and ozone as a starting point, debate the proposition that it would be unreasonable to eliminate all of the environmental risks generated by technology.

141

Chapter seven

The greenhouse effect

The greenhouse effect is one of the best known of current global environmental issues. Media coverage has helped to generate widespread popular interest in the topic and its many ramifications. Concern is also high in the scientific community, where the greenhouse effect is being investigated at all organizational levels and in a wide range of disciplines. Most of the media, and also most of those involved in the investigation and analysis of the greenhouse effect, seem to have accepted its impact as already a *fait accompli*. There are only a few dissenting voices expressing misgivings about the nature of the evidence and the rapidity with which it has been embraced. All admit the need for further research and, as a result of the investigations set in motion in the past several years, it is likely that the greenhouse effect will maintain a high profile as an environmental issue for some time to come.

The creation of the greenhouse effect

The greenhouse effect is brought about by the ability of the atmosphere to be selective in its response to different types of radiation. The atmosphere readily transmits solar radiation – which is mainly short-wave from the ultraviolet end of the energy spectrum – allowing it to pass through unaltered to heat the earth's surface. The energy absorbed by the earth is re-radiated into the atmosphere, but this terrestrial radiation is long-wave infrared, and instead of being transmitted it is absorbed, causing the temperature of the atmosphere to rise. Some of the energy absorbed in the atmosphere is returned to the earth's surface, causing its temperature to rise also (see Chapter 2). This is considered similar to the way in which a greenhouse works – allowing sunlight in but trapping the resulting heat inside – hence the use of the name 'greenhouse effect'. In reality it is the glass in the greenhouse which allows the

142

temperature to be maintained, by preventing the mixing of the warm air inside with the cold air outside. There is no such barrier to mixing in the real atmosphere, and some scientists have suggested that the processes are sufficiently different to preclude the use of the term 'greenhouse effect'. Anthes *et al.* (1980), for example, prefer to use 'atmospheric effect'. However, the use of 'greenhouse effect' to describe the ability of the atmosphere to absorb infrared energy is so well established that any change would cause needless confusion. The demand for change is not strong, and, although the analogy is not perfect, the term 'greenhouse effect' will continue to be used widely for descriptive purposes.

Without the greenhouse effect global temperatures would be much lower than they are, perhaps averaging only $-17\,°C$ compared to the existing average of $+15\,°C$. This, then, is a very important characteristic of the atmosphere, yet it is made possible by a group of gases which together make up less than 1 per cent of the total volume of the atmosphere. There are about twenty of these greenhouse gases. Carbon dioxide is the most abundant, but methane, nitrous oxide, and the chlorofluorocarbons are potentially significant, and water vapour also exhibits greenhouse properties. Any change in the volume of these gases will disrupt the energy flow in the earth/atmosphere system, and this will be reflected in changing world temperatures. This is nothing new. Although the media sometimes seem to suggest that the greenhouse effect is a modern phenomenon, it is not. It has been a characteristic of the atmosphere for millions of years, sometimes more intense than it is now, sometimes less.

The changing greenhouse effect

Palaeoenvironmental evidence suggests that the greenhouse effect fluctuated quite considerably in the past. In the Quaternary era, for example, it was less intense during glacial periods than during the interglacials (Bach 1976, Pisias and Imbrie 1986). Present concern is with its increasing intensity and the associated global warming. The rising concentration of atmospheric CO_2 is usually identified as the main culprit, although it is not the most powerful of the greenhouse gases. It is the most abundant, however, and its concentration is increasing rapidly. As a result, it is considered likely to give a good indication of the trend of the climatic impact of the greenhouse effect, if not its exact magnitude.

Svante Arrhenius, a Swedish chemist, is usually credited with being the first to recognize that an increase in CO_2 would lead to global warming (Bolin 1972, Bach 1976, Crane and Liss 1985).

Other scientists, including John Tyndall in Britain and T.C. Chamberlin in Sweden, also investigated the link, but Arrhenius provided the first quantitative predictions of the rise in temperature (Idso 1981, Crane and Liss 1985). He published his findings at the beginning of this century, at a time when the environmental implications of the Industrial Revolution were just beginning to be appreciated. Little attention was paid to the potential impact of increased levels of CO_2 on the earth's radiation climate for some time after that, however, and the estimates of CO_2-induced temperature increases calculated by Arrhenius in 1903 were not bettered until the early 1960s (Bolin 1972). Occasional papers on the topic appeared (e.g. Callendar 1938, Revelle and Seuss 1957, Bolin 1960), but interest began to increase significantly only in the early 1970s, as part of a growing appreciation of the potentially dire consequences of human interference in the environment. Increased CO_2 production and rising atmospheric turbidity were recognized as two important elements capable of causing changes in climate. The former had the potential to cause greater warming, whereas the latter was considered more likely to cause cooling (Schneider and Mesirow 1976). For a time it semed that the cooling would dominate (Calder 1974, Ponte 1976), but a spate of papers on the greenhouse warming, published in the early 1980s, changed that (e.g. Idso 1980, Manabe *et al.* 1981, Schneider and Thompson 1981, Pittock and Salinger 1982, Mitchell 1983, NRC 1982 and 1983). They revealed that scientists had generally underestimated the speed with which the greenhouse effect was intensifying, and had failed to appreciate the impact of the subsequent global warming on the environment or on human activities.

Worldwide concern, coupled with a sense of urgency uncommon in the scientific community, led to a conference on the 'International Assessment of the Role of Carbon Dioxide and other Greenhouse Gases in Climate Variations and Associated Impact', held at Villach, Austria, in October 1985. To ensure the follow-up of the recommendations of that conference, an Advisory Group on Greenhouse Gases (AGGG) was established under the auspices of the International Council of Scientific Unions (ICSU), the United Nations Environment Programme (UNEP), and the World Meteorological Organization (WMO) (Environment Canada 1986). The main tasks of the AGGG were to carry out biennial reviews of international and regional studies related to the greenhouse gases, to conduct aperiodic assessments of the rates of increases in the concentrations of greenhouse gases, and to estimate the effects of such increases. Beyond this, they also supported further studies of the socio-economic impacts of climatic change produced by the greenhouse gases, and identified areas such as the monsoon region of south-east Asia, the Great Lakes region of North America, and the circumpolar Arctic as likely candidates for increased investigation. The AGGG suggested that the dissemination of information on recent developments to a wide audience was also important, and in keeping with that viewpoint Environment Canada began the production of a regular newsletter to highlight current events in CO_2/climate research. Its annual reports for 1985 and 1986, *Understanding CO_2 and Climate*, were also devoted to that theme. In addition, Environment Canada has funded research on such diverse topics as the effect of climatic change on agriculture in Ontario (Smit 1987), the impact of rising sea-levels on Prince Edward Island (Lane *et al.* 1988), and the effects of CO_2-induced warming on the ski industry in Ontario (Wall 1988), which have ensured that Canada is one of the leaders in the investigation of the environmental and socio-economic impact of global warming. The Department of Energy in the United States has also been active in the field, with more broadly based reports on the effects of increasing CO_2 levels on vegetation (Strain and Cure 1985) and on climate (MacCracken and Luther 1985a and 1985b), as well as the effects of future energy use and technology on the emission of CO_2 (Edmonds *et al.* 1986, Cheng *et al.* 1986).

In Europe, Flohn's (1980) study of the climatic consequences of global warming caused by human activities for the International Institute for Applied Systems Analysis (IIASA) included consideration of CO_2. More recently, the Commission of European Communities (CEC) has funded research into the socio-economic impacts of climate changes which might be caused by a doubling of atmospheric CO_2 (Meinl *et al.* 1984, Santer 1985). Most of these investigations have involved the use of GCMs, and the UK Meteorological Office five-layer GCM has provided information on CO_2-induced climatic change over western Europe (Wilson and Mitchell 1987). Several other European countries, including Germany and The Netherlands, have launched research programmes, but interest in the topic appears less than in North America.

The carbon cycle and the greenhouse effect

Carbon is one of the most common elements in the environment. It is present in all organic substances, and is a constituent of a great variety of compounds, ranging from relatively simple gases to very complex derivatives of petroleum hydrocarbons. Three of the

145

principal greenhouse gases – CO_2, methane (CH_4), and the CFCs – contain carbon. The carbon in the environment is mobile, readily changing its affiliation with other elements in response to biological, chemical, and physical processes. This mobility is controlled through a natural bio-geochemical cycle which works to maintain a balance between the release of carbon compounds from their sources and their absorption in sinks. The natural carbon cycle is normally considered to be self-regulating, but with a time scale of the order of thousands of years. Over shorter periods, the cycle appears to be unbalanced, but that may be a reflection of an incomplete understanding of the processes involved or perhaps an indication of the presence of sinks or reservoirs still to be discovered (Moore and Bolin 1986). The carbon in the system moves between several major reservoirs (see Figure 7.1). The atmosphere, for example, contains more than 700 billion tonnes of carbon at any given time, while 1,800 billion tonnes are stored on land, and close to 40,000 billion tonnes are contained in the oceans (Gribbin 1978). Living terrestrial organic matter is estimated to contain between 450 and 600 billion tonnes (Moore and Bolin 1986). World fossil fuel reserves also constitute an important carbon reservoir of some 5,000 billion tonnes (McCarthy *et al.* 1986). They contain carbon which has not been active in the cycle for millions of years, but is now being reintroduced as a result of the growing demand for energy in modern society being met by the mining and burning of fossil fuels. It is being reactivated in the form of CO_2, which is being released into the atmospheric reservoir in quantities sufficient to disrupt the natural flow of carbon in the environment. The sinks are unable to absorb the CO_2 as rapidly as it is being produced. The excess remains in the atmosphere, to intensify the greenhouse effect, and thus contribute to global warming.

The burning of fossil fuels adds 5 billion tonnes of CO_2 to the atmosphere every year (Keepin *et al.* 1986), and that remains the primary source of anthropogenic CO_2. Augmenting that is the destruction of natural vegetation which causes the level of atmospheric CO_2 to increase by reducing the amount recycled during photosynthesis. Photosynthesis is a process, shared by all green plants, by which solar energy is converted into chemical energy. It involves gaseous exchange. During the process, CO_2 taken in through the plant leaves is broken down into carbon and oxygen. The carbon is retained by the plant while the oxygen is released into the atmosphere. The role of vegetation in controlling CO_2 through photosynthesis is clearly indicated by variations in the levels of the gas during the growing season. Measurements at

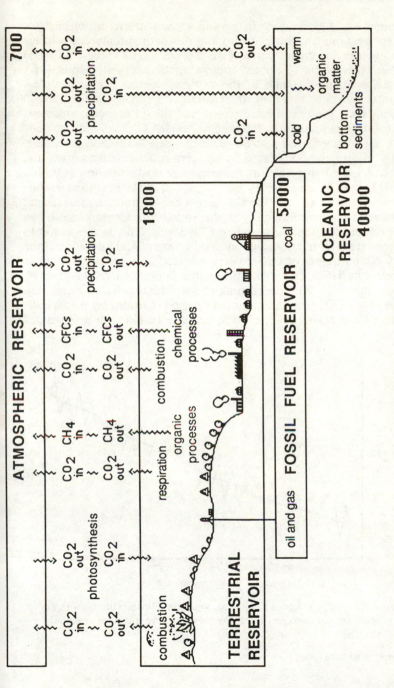

Figure 7.1 Schematic representation of the storage and flow of carbon in the earth/atmosphere system

Source: Compiled from data in Gribbin (1978), McCarthy *et al.* (1986)

Mauna Loa Observatory in Hawaii show patterns in which CO_2 concentrations are lower during the northern summer and higher during the northern winter (see Figure 7.2). These variations reflect the effects of photosynthesis in the northern hemisphere, which contains the bulk of the world's vegetation (Bolin 1986). Plants absorb CO_2 during their summer growing phase, but not during their winter dormant period, and the difference is sufficient to cause semi-annual fluctuations in global CO_2 levels.

Photosynthesis will also be reduced when vegetation is cleared. This is usually considered a modern phenomenon, with the destruction of tropical forests receiving most attention (Gribbin 1981), but Wilson (1978) has suggested that the pioneer agricultural settlement of North America, Australasia, and South Africa in the second half of the nineteenth century made an important contribution to rising CO_2 levels. This is supported to some extent by the observation that between 1850 and 1950 some 120 billion tonnes of carbon were released into the atmosphere as a result of deforestation and the destruction of other vegetation by fire (Stuiver 1978). The burning of fossil fuels produced only half that much CO_2 over the same time period. Current estimates indicate that the atmospheric CO_2 increase resulting from reduced

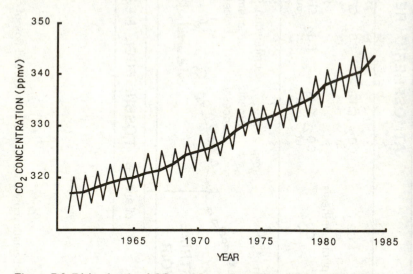

Figure 7.2 Rising levels of CO_2 at Mauna Loa, Hawaii (the smooth curve represents annual average values; the zig-zag curve indicates seasonal fluctuations)

Source: After Bolin (1986)

photosynthesis and the burning of vegetation is equivalent to about 1 billion tonnes per year (Moore and Bolin 1986), down slightly from the earlier value. However, the annual contribution from the burning of fossil fuels is almost ten times what it was in the years between 1850 and 1950.

Although the total annual input of CO_2 to the atmosphere is of the order of 6 billion tonnes, the atmospheric CO_2 level increases by only about 2.5 billion tonnes per year. The difference is distributed to the oceans and other sinks (Moore and Bolin 1986). The oceans absorb 2.5 billion tonnes, some as a result of photosynthesis in phytoplankton, some through nutritional processes which allow marine organisms to grow calcium carbonate shells or skeletons, and some by direct diffusion at the air/ocean interface (McCarthy *et al.* 1986). The mixing of the ocean waters causes the redistribution of the absorbed CO_2. In polar latitudes, for example, the added carbon sinks along with the cold surface waters in that region, whereas in warmer latitudes carbon-rich waters well up towards the surface allowing the CO_2 to escape again. The turnover of the deep ocean waters is relatively slow, however, and carbon carried there in the sinking water or in the skeletons of dead marine organisms remains in storage for hundreds of years. More rapid mixing takes place through surface ocean currents such as the Gulf Stream, but in general the sea responds only slowly to changes in atmospheric CO_2 levels. This may explain the apparent inability of the oceans to absorb more than 40–50 per cent of the CO_2 added to the atmosphere by human activities, although it has the capacity to absorb all of the additional carbon (Moore and Bolin 1986).

The oceans constitute the largest active reservoir of carbon in the earth/atmosphere system, and their ability to absorb CO_2 is not in doubt. However, the specific mechanisms involved are now recognized as extremely complex, requiring more research into the interactions between the atmosphere, ocean, and biosphere if they are to be better understood (Crane and Liss 1985).

Atmospheric carbon dioxide and temperature change

Although present concern with atmospheric CO_2 centres on rising concentrations of the gas, concentrations have varied considerably in the past. Analysis of air bubbles trapped in polar ice indicates that the lowest levels of atmospheric CO_2 occurred during the Quaternary glaciations (Delmas *et al.* 1980). At that time, the atmosphere contained only 180 to 200 parts per million (ppm) of CO_2, although there is some evidence that levels fluctuated by as

149

much as 60 ppm in periods as short as 100 years (Crane and Liss 1985). Levels rose to 275 ppm during the warm inter-glacial phases, and that level is also considered representative of the pre-industrial era of the early nineteenth century (Bolin 1986). CO_2 measurements taken by French scientists in the 1880s, just as the effects of the Industrial Revolution were beginning to be felt, have been reassessed by Siegenthaler (1984), who has concluded that levels in the northern hemisphere averaged 285 to 290 ppm at that time.

When the first measurements were made at the Mauna Loa Observatory in 1957, concentrations had risen to 310 ppm, and they continued to rise by just over 1 ppm per year to reach 335 in 1980. Since then, levels have risen at a rate of 2 to 4 ppm per year to reach the present level of 345 ppm (Gribbin 1981, Bolin 1986). That represents an increase of 70 ppm in less than 200 years. The difference between glacial and inter-glacial periods was about the same, but then the time interval was measured in tens of thousands of years. If the present rate of increase continues until the year 2050, it is estimated that the atmospheric CO_2 concentration will be 450 ppm, and by 2075 it will be 500–600 ppm, more than twice the 1800 level (Bolin 1986) (see Figure 7.3).

These values in turn can be translated into temperature increases. It has been estimated, for example, that a 0.3–1.1 °C increase in the earth's surface temperature has taken place since 1900 because of the greenhouse effect. Schneider (1987) claims that the earth is 0.5° warmer in the 1980s than it was in the 1880s. Such increases may be too small to be detected among the normal natural variations in global temperatures (Crane and Liss 1985), but Hansen and Lebedeff (1988) calculated that the warming between the 1960s and 1980s has been more rapid than that between the 1880s and 1940s, which suggests that the greenhouse warming may be beginning to emerge from the general back-ground 'noise'. Hansen, of the Goddard Institute of Space Studies, claimed subsequently that the global greenhouse signal is sufficiently strong that a cause-and-effect relationship between the CO_2 increase and global warming can be inferred (Climate Institute 1988a).

Estimates of global warming are commonly obtained by employing atmos heric modelling techniques. These have produced a general concensus that a doubling of CO_2 levels would cause an average warming of 1.3–4.5 °C (Manabe and Wetherald 1975, Cess and Potter 1984, Dickinson 1986, Bolin *et al.* 1986). This compares with the estimate of 4–6 °C made by Arrenhius at the beginning of the century (Kellogg 1987). Smaller increases

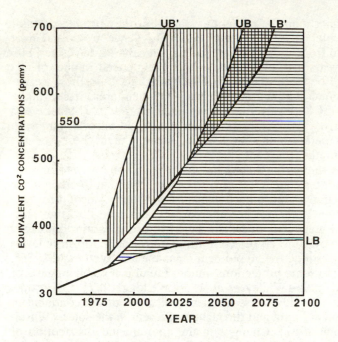

Figure 7.3 Projected changes in atmospheric CO_2 concentration. The curves LB and UB indicate the range of estimates for CO_2 alone. LB′ and UB′ inciate the range of estimates when other greenhouse gases are included and expressed in CO_2 equivalents. The value 550 ppm corresponds to a doubling of pre-industrial concentrations

Source: After Bolin *et al.* (1986)

have been calculated by Newell and Dopplick (1979) who estimated that the temperature above the tropical oceans would increase by only 0.03 °C on average and by Idso (1980) who estimated a global increase of 0.26 °C. These lower values are generally considered to be unrepresentative by most scientists investigating the problem, however (Cess and Potter 1984, Webster 1984).

Although the estimated temperature increases are not particularly impressive – mainly because they are average global values – evidence from past world-temperature changes indicates that they are of a magnitude which could lead to significant changes in climate and climate-related activities. During the climatic optimum – the warm epoch following the last ice age – some 5,000 to 7,000 years ago, temperatures in North America and Europe were only

2–3 °C higher than the present average, but they produced major environmental changes (Lamb 1977). Evidence from that time period, and from another warm spell in the early Middle Ages, 800 to 1,000 years ago, also suggests that the greatest impact of any change will be felt in mid to high latitudes in the northern hemisphere. Actual temperature changes in such areas following a doubling of CO_2 levels might be as much as 5–9 °C (Manabe and Wetherald 1986). Seasonal variations would also occur, with winter temperature increases being greater than those for the summer months (see Table 7.1).

Because the various elements in the atmospheric environment are closely interrelated, it is only to be expected that if temperature changes, changes in other elements will occur also. Moisture patterns are likely to be altered, for example. In the mid-continental grasslands of North America, precipitation totals may be reduced by 10 to 20 per cent, while summer soil moisture levels may fall by as much as 50 per cent (Manabe and Wetherald 1986). According to some projections, more rainfall is possible in parts of Africa and India (Wigley *et al.* 1986, Kellogg 1987). In Europe, precipitation would be less frequent over most of the continent in summer and autumn, and throughout the year in the south (Wilson and Mitchell 1987). Changes in the amount and distribution of rainfall, coupled with changes in the length and intensity of the growing season, would disrupt existing vegetation patterns, and cause major alterations in agricultural activities in many areas.

The reality of the situation may only become apparent when the changes have occurred, for there are many variables in the predictions. The human factors, as always, are particularly unpredictable. Technology, politics, socio-economic conditions, and even demography can contribute to changes in the concentration of CO_2, yet the nature and magnitude of the variations in these elements is almost impossible to predict. As a result, most attempts at projecting changes into the future have dealt with ranges of possibilities rather than discreet values.

One common approach, aimed at simplifying the human factors as much as possible, has been to compress the wide selection of variables into one element which combines energy efficiency and fossil fuel use (Bolin *et al.* 1986). The rationale behind this is the fact that the primary source of anthropogenic CO_2 is the burning of fossil fuels for the production or release of energy (Gribbin 1981). It is estimated that such processes annually emit some 5 billion tonnes of carbon, in the form of carbon dioxide, into the atmosphere. On the assumption that such factors as technological levels in society and economic development are related to the

Table 7.1 Projected temperature changes at selected locations in the northern hemisphere following a doubling of atmospheric CO_2

	Mean temperatures (°C)					
	Winter present	Winter $2 \times CO_2$	Increase	Summer present	Summer $2 \times CO_2$	Increase
Calgary, Alberta: 51° N	−8.3	−6.3	2	14.9	16.9	2
Toronto, Ontario: 43° N	−3.3	−0.3	3	20.7	23.7	3
Angmagssalik, Greenland: 65° N	−6.3	1.7	8	6.7	7.7	1
Anchorage, Alaska: 61° N	−10.2	−1.2	9	13.5	17.5	4
San Francisco, California: 37° N	9.5	11.5	2	15.5	18.5	3
Chicago, Illinois: 41° N	−2.3	0.7	3	22.7	25.7	3
Reykyavik, Iceland: 64° N	−0.2	9.8	10	10.5	12.5	2
Edinburgh, Scotland: 56° N	4.0	12.0	8	14.0	18.0	4
Paris, France: 48° N	3.4	8.4	5	18.0	20.0	2
Oslo, Norway: 59° N	−3.7	5.3	9	16.0	20.0	4
Leningrad, USSR: 59° N	−6.7	3.3	10	16.7	19.7	3
Vladivostok, USSR: 43° N	−11.3	−5.3	6	18.3	20.3	2
Beijing, China: 39° N	−3.3	1.7	5	25.3	26.3	1

Source: Estimated from data in Manabe and Wetherald (1986), Washington and Meehl (1984)

demand for energy, the latter can be used as a surrogate for the other factors. The most recent estimates of the magnitude and timing of changes in the greenhouse effect fall within a range delimited by a combination of energy efficiency and fossil-fuel use. At one end of the scale is the 'high energy efficiency/low fossil-fuel use' scenario in which carbon emissions fall to 2 billion tonnes per year. At the other end of the scale is the situation in which fossil-fuel use will remain high and energy efficiency will be low, allowing as much as 20 billion tonnes of carbon per year to be released into the atmosphere (Keepin *et al.* 1986). The exact position of future emissions within this range is far from clear, but in terms of atmospheric CO_2 levels it could mean that a doubling of pre-industrial concentrations will occur as early as the year 2050 (low energy efficiency/high fossil-fuel use) or as late as the middle of the twenty-second century (high energy efficiency/low fossil-fuel use) (see Figure 7.3). This in turn will have a significant effect on the extent and timing of the global temperature change.

The contribution of other greenhouse gases

Most predictions of future changes in the intensity of the greenhouse effect are based solely on changes in the CO_2 content of the atmosphere. Their accuracy is therefore questionable, since CO_2 is not the only greenhouse gas, nor is it the most powerful. Methane (CH_4), nitrous oxide (N_2O), and the chlorofluorocarbons (CFCs) are the most important of the other greenhouse gases. Tropospheric ozone (O_3) is also capable of enhancing the greenhouse effect, but its present concentrations are very variable in both time and place, and there is no clear indication of future trends (Bolle *et al.* 1986).

Levels of atmospheric CH_4 are very low compared to those of CO_2. It is about three times more effective than CO_2, however, and its concentration is increasing at about 1.1 per cent to 1.3 per cent per year (Bolle *et al.* 1986). The most important cause of this increase is the world's growing population of domestic animals such as cattle, pigs, and sheep, which release considerable amounts of CH_4 through their digestive systems. Significant amounts of CH_4 are also produced by the global termite population, and the increasing land area devoted to rice paddies contributes to rising levels (Crutzen *et al.* 1986).

The current atmospheric concentration of N_2O is close to that of CO_2, but is increasing less rapidly than either CO_2 or CH_4. N_2O is produced naturally in the environment through nitrification, but it owes its present growth to the increased use of fossil fuels and the denitrification of agricultural fertilizers (Chapter 6). The global N_2O budget remains poorly understood, and its future concentration is therefore difficult to predict (Bolle *et al.* 1986).

CFCs released from refrigeration units, insulating foams, aerosol spray cans, and industrial plants are recognized for their ability to destroy the stratospheric ozone layer, but they are also powerful greenhouse gases. Recent international agreements to reduce the use of CFCs are aimed at preventing further damage to the ozone layer, but they will also have some impact on the greenhouse effect. However, CFCs have a long residence time in the atmosphere, and, even as emission rates fall, they will continue to contribute to global warming for some time to come.

The presence of these other greenhouse gases introduces a number of uncertainties into the predictions of future greenhouse levels. None of them is individually as important as CO_2. It has been suggested, however, that their combined influence on the greenhouse effect is already equivalent to half that of CO_2 alone (Bolle *et al.* 1986), and by early next century their contribution to global warming could be equal to that of CO_2 (Ramanathan *et al.* 1985). Their impact would become increasingly important in the low CO_2 emission scenarios envisioned by some investigators (see Figure 7.3). The involvement of the CFCs and N_2O in the depletion of the ozone layer adds a further complication. Attempts to mitigate the effects of these gases on the ozone layer would also impact on the greenhouse effect. Thus, although such gases as CH_4, N_2O, and the CFCs have received much less attention than CO_2 in the past, it is clear that plans developed to deal with global warming must include consideration of all the greenhouse gases, not just CO_2.

Environmental and socio-economic impacts of increasing greenhouse gases

Given the wide range of possibilities presented in the estimates of future greenhouse gas levels and the associated global warming, it is difficult to predict the environmental and socio-economic effects of such developments. However, using a combination of investigative techniques – ranging from laboratory experiments with plants to the creation of computer-generated models of the atmosphere and the analysis of past climate anomalies – researchers have produced results which provide a general indication of what the consequences might be in certain key sectors.

The impact of elevated CO_2 levels on natural and cultivated vegetation has received considerable attention. Two elements are involved, since elevated CO_2 has an effect on both photosynthesis and on temperature. Through its participation in photosynthesis, CO_2 provides the carbon necessary for proper plant growth. In laboratory experiments under controlled conditions, it has been shown that increased concentrations of the gas enhance growth in most plants. A doubling of CO_2 has increased yields of maize, sorghum, millet, and sugar cane by 10 per cent in controlled experiments, and increases of as much as 50 per cent have been achieved with temperate zone plants (Environment Canada 1986). The response of natural vegetation or field crops might be less, because of a variety of non-climatic variables, but some trees do seem capable of responding quite dramatically. For example, Sveinbjornsson (1984) has estimated that a doubling of CO_2 would double the rate of photosynthesis in the Alaskan paper birch. Some investigators consider the present levels of CO_2 to be suboptimal for photosynthesis and primary productivity in the majority of terrestrial plants. They suggest, therefore, that the biological effects of enhanced CO_2 would likely be beneficial for most plants (Wittwer 1984).

The predicted higher temperatures, working through the lengthening and intensification of the growing season, would have an effect on the rates of plant growth and crop yields. The regional distribution of vegetation would change, particularly in high latitudes where the temperature increases are expected to be greatest (Shugart *et al.* 1986). Across the northern regions of Canada, Scandinavia, and the USSR, the trees of the boreal forest would begin to colonize the tundra, as they have done during warmer spells in the past (Viereck and Van Cleve 1984, Ball 1986). The southern limit of the boreal forest would also migrate northwards, under pressure from the species of the hardwood forests and grasslands which are more suited to the new conditions. That would have a significant effect on countries such as Canada, Sweden, Finland, and the USSR, where national and regional economies depend very much on the harvesting of softwoods from the boreal forest.

The regional distribution of cropland, and the type of crop grown on that land, would change as temperatures rise. A significant expansion of agriculture is to be expected in mid to high latitudes, where the greatest warming will be experienced. In the interior of Alaska, for example, a doubling of CO_2 levels would raise temperatures sufficiently to lengthen the growing season by 3 weeks (Wittwer 1984), which would allow land presently under forage crops, or even uncultivated land, to produce food crops such as cabbage, broccoli, carrots, and peas. The growing season in Ontario, Canada, would be lengthened by 48 days in the north and 61 in the south. With the longer season, corn, wheat, and soybeans would become viable northern Ontario crops (Smit 1987). The greater intensity of the growing season, along with the effects of increased CO_2 on photosynthesis, would lead to increased crop yields. In Europe, simulations of the effects of a doubling of CO_2 on crops, using grass as a reference, have indicated an average increase in biomass potential of 9 per cent, with regional values ranging from an increase of 36 per cent in Denmark to a decrease of 31 per cent in Greece (Santer 1985). Similarly, projected improvements in the growing season in Manitoba, Canada, have the potential to produce a 15 per cent increase in crop yields (Environment Canada 1986).

It may not always be possible for agriculturalists to take advantage of the benefits of global warming, because of the effects of non-climatic elements on agricultural production. Warmer climates would allow the northward expansion of cultivation on the Canadian prairies, for example, but the benefits of that would be offset by the inability of the soils in those areas to support anything other than marginal forage crops, which are not profitable under current economic conditions (Arthur 1988). Few simulations take such variables into account. The model employed by Santer (1985) to predict changes in European biomass potential included no provision for such important non-climatic elements as insects, disease, and additional fertilizers. Until such unknowns can be estimated, the results of this and similar simulations must be treated with caution.

The most serious problem associated with the warming is a climatological one. It is the increasing dryness likely to accompany the rising temperatures in many areas. Reduced precipitation following changes in circulation patterns, plus the increased rates of evapotranspiration caused by higher temperatures, would create severe moisture stress for crops in many areas (Climate Institute 1988b). Less precipitation and higher temperatures in the farmlands of southern Ontario might reduce yields sufficiently to cause losses of as much as $100 million per year (Smit 1987). The areas hardest hit would be the world's grain-producing areas, which would become drier than they are now, following the global warming (Kellogg 1987). Corn yields would be reduced in the midwestern plains of the United States, and a major increase in the frequency and severity of drought would lead to more frequent crop failures in the wheat-growing areas of Canada (Williams *et al.* 1988). The grain belt in the USSR, already unable to meet that nation's needs, would suffer as badly as its North American equivalent (Kellogg 1987).

Such developments would disrupt the pattern of the world's grain trade, which depends heavily on the annual North American surplus. Food supply problems would become serious in the USSR, and famine would strike many Third World countries. The picture is not completely hopeless, however, for rainfall is expected to increase in some tropical areas, and the combination of more rain, high temperatures, and more efficient photosynthesis could lead to increased rice yields of as much as 10 per cent (Gribbin 1981). Some of the grain-growing areas in Australia might also experience increased precipitation and higher temperatures (Kellogg 1987).

Most of the agricultural projections note the importance of an adequate water supply if the full benefits of global warming are to be experienced. Beyond that, little work has been done on the impact of an intensified greenhouse effect on water resources. Canadian studies have examined the implications of climate change for future water resources in the Great Lakes–St Lawrence River system (Sanderson 1987), but the general lack of attention to the hydrologic cycle following global warming is an important gap in current studies.

One aspect of global hydrology which has been considered in some detail is the impact of higher world temperatures on sea level. It is possible that as early as the year 2050, mean sea level will have risen by as much as 1 m, as a result of the thermal expansion of the oceans and the melting of temperate glaciers (Titus 1986). Although this is relatively minor compared to past changes in sea level, it would be sufficient to cause serious flooding and erosion in coastal areas. In low-lying regions such as The Netherlands, which are already dependent upon major protective works, even a sea-level rise of half a metre would have major consequences (Hekstra 1986). Land close to sea level in Britain – around the Wash, for example – would be similarly vulnerable, and structures like the Thames Flood Barrier might be needed on other British rivers. Environment Canada has commissioned studies of the impact of sea-level rises in the Maritime Provinces, which show that flooding events or storm surges would become more frequent and severe, presenting serious problems for sewage and industrial waste facilities, road and rail systems, and harbour activities (Martec Ltd 1987, Stokoe 1988). They also indicate that the relationship between sea level and regional hydrology would allow the effects to be felt some distance inland in the form of altered streamflow patterns and groundwater levels. In time, the impact of higher temperatures on the rate of ablation of ice caps and sea ice in polar regions might be sufficient to cause a rise in mean sea level of 3–4 m (Hoffman *et al.* 1983). Should this ever come to pass, most of the world's major ports would not survive without extensive and costly protection.

This disruption of maritime commercial activities by rising sea level is only one of a series of economic impacts which would accompany global warming. Others range from changes in energy use, particularly for space heating, to the development of entirely new patterns of recreational activity. Because of its northern location, Canada is especially susceptible to such changes, and as a result Environment Canada has invested a great deal of time and effort in viewing the greenhouse effect from a Canadian perspective (see Figure 7.4). The results of the studies, although specific to Canada, may also give an indication of future developments in other northern regions – such as Scandinavia and the USSR (Environment Canada 1987).

Alternative points of view

The global warming scenario, in which a doubling of atmospheric CO_2 would cause an average temperature increase of 1.3–4.5 °C, is widely accepted. This concensus has evolved from the results of many experiments with theoretical climate models (Dickinson 1986). These models represent the earth's physical processes through a series of mathematical formulae, which may be as simple or as complex as required. The most sophisticated models currently in use are the 'general circulation models' (GCMs), which are capable of representing changing climatic characteristics by latitude, longitude, and altitude (Schneider 1987). To examine

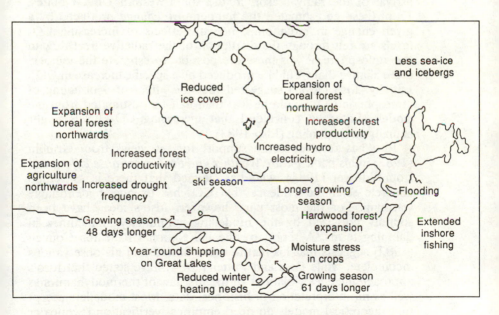

Figure 7.4 Environmental and economic changes expected in central and eastern Canada following a doubling of atmospheric CO_2 levels

the impact of an enhanced greenhouse effect, the CO_2 component in the model is increased to a specific level – usually 2× or 4× – and the model is allowed to run until equilibrium is established among the various climatic elements included. New temperature values are most commonly predicted in this way, but any element affected by rising CO_2 levels can be considered. However, even the most complex models cannot exactly represent the working of the atmosphere. Those who use the models are often the first to note their limitations, but most of the criticisms of the estimated impact of elevated CO_2 levels on climate have arisen out of the perceived inadequacies of the models used.

The most vociferous critic of the theoretical-modelling approach is Sherwood Idso, who has attacked the established view of future global warming through numerous publications (e.g. Idso 1980, 1981, 1982, 1987). He has suggested that increasing CO_2 levels would produce negligible warming, and might even cause global cooling. His conclusions were based on so-called natural experiments. He monitored temperature change and radiative heat flow during natural events such as dust storms, and measured the build-up of atmospheric moisture levels with the arrival of the rainy season in the south-western United States. From these he estimated the temperature change produced by a given change in radiation. Since the effects of increasing CO_2 levels are felt through the disruption of the radiative heat flow in the atmosphere, it was therefore possible to estimate the temperature change that would be produced by a specific increase in CO_2. Initially, Idso (1980) suggested that the effect of a doubling of atmospheric CO_2 would be less than half that estimated from the models. Later he concluded that increasing CO_2 levels might actually cause cooling (Idso 1983).

Idso has received some support for his views from Gribbin (1982) in his book *Future Weather and the Greenhouse Effect*, and from Wittwer (1984), who has claimed that there is as yet no concrete scientific evidence that CO_2 has produced identifiable warming. For the most part, however, Idso's ideas have been soundly criticized by the modelling community, sometimes in damning terms. A report by the US National Research Council (1982) suggests that his methods are flawed and his observations incomplete, while Cess and Potter (1984) have stated that Idso's approach includes violations of the first law of thermodynamics as well as misinterpretation of historical data. Most modellers accept that theoretical models do need empirical verification (Schneider 1984), and Potter *et al.* (1987) have suggested that Idso's natural experiments are not inconsistent with the modelling approach.

They claim that the main inconsistency arises from Idso's interpretation of the results of his natural experiments.

The two groups remain polarized over their conclusions, but the
debate continues. At times, it has been acrimonious and
emotional. Its intensity has tended to obscure other aspects of the
issue, which have therefore received less attention than they
deserve.

The concentration on one variable, CO_2 levels, has allowed the
role of other elements in the earth/atmosphere system to be
largely ignored. There are natural processes which could augment
the impact of increased CO_2 and there are others which could
reduce it. Human-induced changes could also ruin the projections.

It is well known that the earth's climate is not static, but has
varied over the years (see e.g. Lamb 1977). Some of the variations
have been major, such as the Ice Ages, whereas others have been
detectable only through detailed instrumental analysis. Some have
lasted for centuries, some for only a few years. While it is relatively
easy to establish that climatic change has taken place, it is quite
another matter to identify the causes. There are a number of
elements considered likely to contribute to climatic change, however.

Since the earth/atmosphere system receives the bulk of its
energy from the sun, any variation in the output of solar radiation
has the potential to cause the climatic change. The links between
sun-spot cycles and changes in weather and climate have long been
explored by climatologists (see e.g. Lamb 1977). Even if the solar-
energy output remains the same, changes in earth–sun
relationships may alter the amount of radiation intercepted by the
earth. Variations in the shape of the earth's orbit, or the tilt of its
axis, for example, have been implicated in the development of the
Quaternary glaciations (Pisias and Imbrie 1986). The present
intensification of the greenhouse effect is directly linked to the
anthropogenic production of CO_2; in the past, however, CO_2
levels have increased with no human contribution whatsoever.
During the Cretaceous period, millions of years before the Industrial Revolution, CO_2 concentrations were much higher than they
are today (Schneider 1987). Other changes in the composition and
circulation of the atmosphere have to be considered also. The
impact of increased atmospheric turbidity is not clear (see Chapter
5). It may add to the general warming of the atmosphere (Bach
1976) but it has also been used to explain global cooling between
1940 and 1960, at a time when CO_2 levels continued to rise (see
Figure 7.5). Although this cooling is usually acknowledged as a
problem by researchers, it has not yet been adequately explained
(Wigley *et al.* 1986).

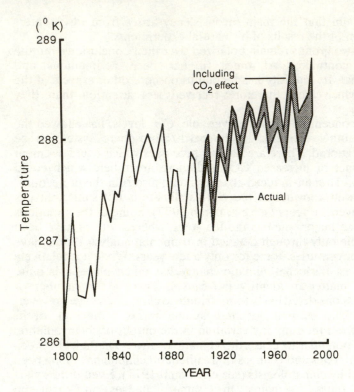

Figure 7.5 Changing global annual surface temperature

Source: After Schneider and Mass (1975)
Note: The lower line indicates actual change, while the upper line indicates the temperatures to be expected if the enhanced greenhouse effect is included

Thus, there are many elements in the earth/atmosphere system capable of producing measurable climatic change. Is it feasible to expect that they will remain dormant while anthropogenic CO_2 provides its input?

Another aspect of the issue which requires additional attention is the ability of feedback mechanisms to increase or decrease the impact of the greenhouse effect. For example, the higher temperatures associated with an intensified greenhouse effect would bring about more evaporation from the earth's surface. This would lead to increased cloudiness as the rising water vapour condensed. The clouds in turn would reduce the amount of solar radiation reaching the surface, and therefore cause a temperature reduction, which might moderate the increase caused by the greenhouse effect.

Unfortunately, general circulation models have great difficulty in dealing with cloudiness (Henderson-Sellers 1986) and its impact is difficult to predict.

The increased cloudiness is considered a negative feedback mechanism since it tends to counter the effects of the original temperature change. There are also positive feedback mechanisms which augment the change. The colder northern waters of the world's oceans, for example, act as an important sink for CO_2, but their ability to absorb the gas decreases as temperatures rise (Bolin 1986). However, global warming (which is expected to be significant in high latitudes) would reduce the ability of the oceans to act as a sink. Instead of being absorbed by the oceans, CO_2 would remain in the atmosphere, thereby adding to the greenhouse effect. The role of feedback mechanisms is important, but current modelling techniques are generally inadequate to deal with them. Such constraints in the existing models must be recognized, and appropriate allowances made when predictions of global warming are used.

There can be no simple response to the changes likely to follow the intensification of the greenhouse effect. Since the present elevated CO_2 levels are directly linked to the increased use of fossil fuels and the destruction of natural vegetation, most suggestions for dealing with the problem include consideration of these two elements. Neither can be altered sufficiently to cancel or reverse the warming trend, however. Even under the 'high energy efficiency/ low fossil-fuel use' scenario, the doubling of atmospheric CO_2 levels will ultimately take place (see Figure 7.3). If the warming cannot be prevented, it must somehow be managed. This would include minimizing the negative effects (such as increased drought or rising sea levels) but it should also involve the maximization of the positive effects (such as the extended growing season or the greater precipitation in Africa and Asia). The more obvious effects may still be 50 to 60 years away, and as yet little direct planning for future warming has been done. Planners tend to assume that the intensity and direction of the change will continue as projected, but the impact of natural processes could change that, and there is no guarantee that plans made today will be sufficient – or even necessary – 50 years on. Research into global warming is continuing at a high level, and it is possible that a better understanding of its interaction with other elements in the earth/atmosphere system will emerge to resolve some of the existing uncertainties. If not, society will have to deal with future environmental changes in much the same way as it has done in the past – by reacting to them after they have happened.

Summary

Since the latter part of the nineteenth century, the earth's greenhouse effect has been intensifying, largely as a result of human activities. The increased use of fossil fuels has raised the level of CO_2 in the atmosphere, and the destruction of natural vegetation has prevented the environment from restoring the balance. Levels of other greenhouse gases – including CH_4, N_2O, and CFC – have also been rising and the net result has been a gradual global warming. If present trends continue, a rise in mean global temperatures of between 1.3 °C and 4.5 °C is projected by the early part of the twenty-first century. These values are not accepted by all scientists studying the situation, but there is enough evidence to suggest that a major global climate change is in progress. The ultimate magnitude of the change is uncertain, but it has the potential to cause large-scale alterations to the natural environment and to global socio-economic and political systems. For these reasons, the search for a better understanding of the situation is being given priority, both nationally and internationally. There has always been a high degree of co-operation among the world's environmental scientists, but it seems likely that such co-operation will have to extend to other physical scientists, social scientists, and decision-makers, if society is to be in the best possible position to cope with what could well be the greatest global temperature rise in history.

Suggestions for further study

1. The general concensus among climatologists is that a doubling of CO_2 will cause a rise in global temperatures of 1.3–4.5 °C. The greater increases are expected to occur at higher latitudes. Choose two high-latitude locations, one maritime and one continental, (e.g. Anchorage and Edmonton or Edinburgh and Moscow) and obtain a listing of their mean monthly temperatures. Assume that following global warming each of the stations experiences an increase of 4.5 °C in its mean annual temperature, the increase being spread evenly through the year. Assume that the mean monthly temperature represents the temperature of the fourteenth day of February, the fifteenth day of a 30-day month, or the sixteenth day of a 31-day month, and assume that the temperature change from one month to the next takes place in regular daily increments. Estimate the following for each station:
   (a) the length of the present growing season in days using 5 °C as a base;

(b) the length of the growing season following global warming;

(c) the length of the present frost-free season in days using 0 °C as a base;

(d) the length of the frost-free season following global warming.

Using that information, write a short account of the differences that would be produced at each station and between stations by the warming. Find out how to estimate changes in the intensity of the growing season from these results, and suggest how agricultural activities might be altered by the warming.

2. Global warming resulting from the intensification of the greenhouse effect is considered capable of causing sea level to rise by about 1 m. For this exercise, find a map of a low-lying coastal or estuarine area at sufficiently large scale that contour intervals of 1 m can be interpolated. From the map, estimate the land area that would be drowned by a 1-metre rise in sea level. What physical features would be destroyed? What human features would be damaged? What might be done to prevent this type of damage? Some estimates have sea level rising ultimately by as much as 5 m, as the water presently stored in glaciers and ice sheets is returned to the sea. Describe the changes which would follow that event in the area covered by your map. Check the altitude of the world's major cities. How many would survive a 5-m rise in sea level? Would it be possible to prepare defences against such an event?

3. Debate the resolution that:

Concentration on negative effects of global warming has obscured the existence of effects which could be potentially beneficial, given appropriate management and planning.

Chapter 8

Nuclear winter

Despite recent agreements between the superpowers to limit the spread of nuclear weapons, the spectre of nuclear war continues to hover over the world, as it has done for the past 40 years or so. Terms such as thermonuclear device, first strike, fallout, and ionizing radiation have become part of the modern lexicon. Added to these now is 'nuclear winter', perhaps the final blow for any survivors of a nuclear exchange (see Figure 8.1).

Environmental problems such as acid rain, ozone depletion, and atmospheric turbidity would be intensified significantly by the multiple nuclear explosions which are a prerequisite for nuclear winter. Nuclear winter differs from these other issues, however. It is a potential problem rather than an existing one, and, as a result, provides no elements capable of direct measurement. It has been necessary, therefore, to develop a theoretical approach using statistical and computerized modelling techniques.

Nuclear winter also differs from other issues in that its effects are catastrophic rather than gradual. The effects of acid rain or ozone depletion may become apparent after only years or even decades; with nuclear winter the effects are felt within days or, at most, weeks of the initial explosions. Thus there is little time to respond once the nuclear devices have been launched. If nuclear winter is to be avoided, it is important that nuclear conflict be avoided also. As a result of recent political developments, however, that conflict seems less likely, and the issue of nuclear winter is regarded as irrelevant by some. Nevertheless, the nuclear powers retain sufficient warheads to create nuclear winter many times over. A broad understanding of the potentially dire consequences of nuclear winter may help to ensure that they are never used.

The pace of research into the causes and effects of nuclear winter is less frenetic than it was between 1983 and 1985 when the issue was first explored in detail. Much of the continuing research

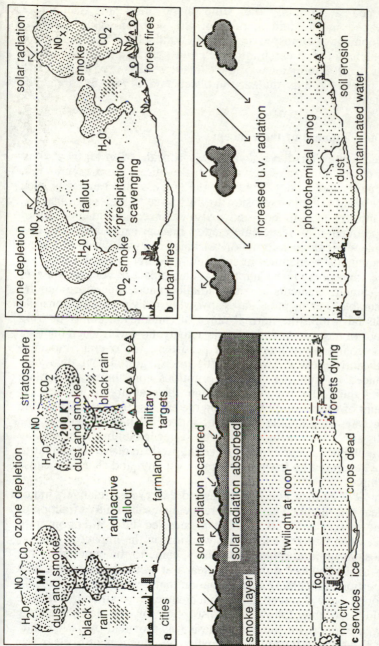

Figure 8.1 The development of nuclear winter: (a) the conflict; (b) post-conflict fires; (c) nuclear winter; (d) the after-effects

is aimed at refining the atmospheric modelling techniques developed in the initial study of the problem. This technical development of atmospheric circulation modelling has been one of the hallmarks of research into nuclear winter, and it has the added bonus that the refinement of the climatic models used in the study may provide results that are applicable to other environmental problems.

The development of the concept

Nuclear explosions have always been considered in terms of their most obvious effects – such as blast, fire, direct ionizing radiation, and radioactive fallout (Peterson 1983). Less direct effects – such as the ability of the explosion to hurl large volumes of dust and soot into the atmosphere – have also been recognized. Because of this production of particulate matter, nuclear explosions are often considered analogous to major volcanic eruptions which also eject contaminants, sometimes in sufficient quantity to form a dust veil in the atmosphere (see Chapter 5). The impact of volcanic dust on such climatic elements as solar radiation and temperature may be measurable, but in most cases involving individual eruptions the results fall within the normal range of climatic variability. During periods of more frequent volcanic activity, however, when the dust veil becomes more intense and persistent, the overall impact is significantly greater. Lamb (1970) has suggested that dust veils produced by multiple volcanic eruptions contributed to cooling during the Little Ice Age, through their ability to reduce the amount of solar radiation reaching the earth's surface. In much the same way, the climatic impact of individual nuclear tests in the 1950s and 1960s seems to have been negligible, but it is possible that multiple nuclear explosions could produce dust veils similar to those generated by frequent volcanic activity, and therefore cause cooling.

In 1977 Ehrlich *et al.* hypothesized that even a relatively minor nuclear war might be capable of reducing insolation to a fraction of its normal value. This possibility received little attention until 1982, when Crutzen and Birks, working on the impact of nuclear explosions on atmospheric chemistry, came to the conclusion that the smoke produced during a nuclear war would block such a high proportion of incoming solar radiation that it would create 'twilight at noon'. They also recognized that changes in weather and climate would inevitably follow this disruption of the earth's energy budget. Support for their hypothesis was provided by Turco *et al.* (1983) who investigated the climatic consequences of nuclear war using a series of physical models originally developed to study the effects of volcanic eruptions. The models indicated a very intense and rapid cooling of the atmosphere in a matter of weeks following even a limited nuclear conflict. The term 'nuclear winter' was coined to describe this cooling, and the work became known as the TTAPS Study, from the initials of the investigators, Turco, Toon, Ackerman, Pollack, and Sagan.

The TTAPS scenario

The nuclear winter hypothesis was based on the assumption that smoke and dust thrown into the atmosphere during a nuclear war would increase atmospheric turbidity to such an extent that a high proportion of incoming solar radiation would be prevented from reaching the lower atmosphere and the earth's surface. The net effect would be to drive land temperatures down (Turco *et al.* 1983). The study employed a one-dimensional radiative convective model (RCM) to estimate the changing vertical distribution of atmospheric smoke and dust particles following a nuclear conflict. Consideration of the impact of these changes on the radiative properties of the atmosphere ultimately allowed air temperatures to be calculated. The model was not particularly sophisticated, however. It could not represent the combined thermal effects of land and water; nor could it forecast short-term or local effects with any accuracy. In addition, the relative simplicity of the RCM prevented it from dealing with the dynamic nature of the atmosphere. For example, changes in global circulation patterns induced by the injection of large amounts of smoke and dust into the atmosphere could not be accommodated in the model, and were not reflected in the results. The TTAPS investigators acknowledged shortcomings such as these, but claimed that the system was sufficient to meet their objective, which was to predict the first-order climate responses of the atmosphere to the introduction of large volumes of particulate matter.

The explosive power of nuclear weapons is immense compared to that of conventional weapons, and this is made apparent by the classification of nuclear devices in terms of their equivalent in conventional high explosives. The atomic bomb dropped on Hiroshima in 1945, for example, had an explosive power equivalent to 15,000 tons of TNT (Bach 1986). This is usually shortened to 15 kilotons, and larger devices may be measured in millions of tons (or megatons) of TNT. The world arsenal of nuclear devices is estimated to contain an explosive power of about 15,000 megatons (Bach 1986). The weapons take various forms from artillery shells to multiple warheads, capable of being delivered by howitzers, aircraft, or missiles. They may be employed strategically against military installations and urban/industrial complexes or used tactically on the battlefield. The ultimate impact of a nuclear war would depend upon the proportion of the total stockpile used, and the way in which the weapons were targeted and delivered. The TTAPS study included a number of cases in which different combinations of these variables were introduced, but the baseline scenario involved a 5,000-megaton war with priority given to military targets, and only about 25 per cent of the total explosive yield assigned to urban or industrial areas (Turco *et al.* 1983).

The initial effect of a nuclear ground-burst would be the injection of large amounts of dust into the atmosphere, through the destruction of soil aggregates, the vaporization of soil and rock, and the incorporation of existing surface dust. Simulations using the TTAPs baseline scenario resulted in an injection of more than 960 million tons of dust into the atmosphere, of which 80 per cent would reach the stratosphere. Such fine dust particles convected into the upper atmosphere would remain in place for more than a year, contributing to global cooling by scattering incoming solar radiation (see Figure 8.1a).

The almost instantaneous injection of dust particles into the atmosphere would be augmented by the more prolonged emission of smoke and soot particles from the massive fires which would follow the initial explosions. These would burn for several weeks in urban areas and in forests, perhaps eventually consuming the equivalent of all of the combustible material in an area of over 750,000 sq. km, adding 225 million tons of smoke to the atmosphere. Urban areas, with their abundance of smoke-producing wood, paper, plastics, and fossil fuels would be particularly potent sources of smoke. According to TTAPS, even a relatively small nuclear exchange of 100 megatons could produce sufficient smoke to cause sub-freezing temperatures in summer, if the attacks were concentrated on cities. Incorporated into the atmospheric circulation, the soot and smoke would first be spread over the northern hemisphere – where it is assumed that most of the nuclear explosions would take place – in a matter of weeks, before being carried into the southern hemisphere, eventually to blanket the entire globe (see Figure 8.1b).

The net result of this rapid increase in atmospheric turbidity would be a severe reduction in the amount of solar radiation reaching the earth's surface. In the baseline case, as little as 3–5 per cent of the incoming solar energy would be able to penetrate the layers of dust and smoke. Such a situation might persist for several weeks until the emission of new aerosols was reduced, and the processes of coagulation, sedimentation, and wet deposition had begun to remove some of the particles. The most visible effect of this would be darkness (perhaps even at noon in mid-latitudes in the nothern hemisphere) or at best a gloomy persistent twilight during the daytime hours. The most direct climatological effect would be a rapid onset of sub-freezing temperatures, whatever the season. Within 3 weeks of the outbreak of nuclear war, average temperatures in the continental interiors – the areas most susceptible to such change – would have fallen to as low as -15–-25 °C. Recovery to a pre-existing average ambient temperature of about 12 °C might take more than a year. The term 'nuclear winter' appears particularly appropriate to describe such circumstances.

Continental interiors would suffer the greatest temperature reductions, with less dramatic changes over the oceans and in coastal areas. Because of the large heat capacity of the oceans, the supply of stored energy would be sufficient to maintain temperatures only a little below normal — perhaps 1–3 °C — even when insolation levels were drastically reduced. The juxtaposition of these relatively mild conditions over the oceans with the cold of the land-masses would create very strong horizontal temperature and pressure gradients. Together these might induce an airflow similar to the present Asiatic winter monsoon, bringing cold air into tropical latitudes, and perhaps more importantly, carrying smoke and dust into the southern hemisphere.

In addition to such variations at the earth's surface, there would be changes in the vertical temperature structure of the atmosphere. Some of the solar energy arriving at the outer edge of a smoky, dust-filled atmosphere would be reflected back into space, but a high proportion would be absorbed, causing temperatures to increase in the lower stratosphere and upper troposphere (see Figure 8.1c). In combination with the falling temperatures in the lower atmosphere, this would intensify the existing stratospheric temperature inversion, and push it down into the upper troposphere. The effects of a long-term inversion of hemispheric proportions are difficult to predict. Its suppression of convective activity would hinder the dispersal of the pollutants. It would also probably lead to a reduction in precipitation, despite the abundance of water vapour and condensation nuclei following the explosions and fires. The intensity of the inversion plus the persistence of the pollution would also cause major changes in atmospheric circulation patterns. The absorption of large amounts of solar radiation in the upper atmosphere would cause the circulation to be driven from above rather than below, as it is at present, and the existing global circulation would be much modified. Changes in regional climatology would follow, but the TTAPS model was incapable of estimating such smaller-scale effects.

The particulates released into the atmosphere by the nuclear explosions and the subsequent fires would be accompanied by a variety of other pollutants. From a climatological point of view, the most important of these would be CO_2 and ozone-destroying compounds such as NO_x. The CO_2 would augment the greenhouse effect, but in the TTAPS study the greenhouse effect was no longer considered to operate following the conflict, and the CO_2 input from nuclear fires was not considered to be climatologically significant. An average ozone reduction of about 30 per cent was included in the baseline study calculations, but its impact on climate was minor compared with that of the dust and smoke.

The climatological disruption following a nuclear exchange is only one of a whole series of problems which would be faced by survivors of the blast, fire, and radiation of the exchange itself (see Figure 8.1d). It might be considered one of the most important, however, since, in addition to its direct effects, its influence would be felt through the workings of natural ecosystems, and those ecosystems modified by society for agricultural use. The potential biological impact of nuclear winter was noted in the TTAPS scenario, but was not elaborated.

Critical analysis and re-assessment of the nuclear winter hypothesis

Depressing and final as it may seem, the TTAPS scenario was no more than a first attempt to assess the impact of nuclear war on world climates. Responses to the nuclear winter hypothesis fall into three main groups. A number of investigators have been highly critical of the approach, for its lack of scientific rigour (Singer 1984, Teller 1984), for the political stance it seems to support (Teller 1984), and for its lack of sensitivity towards the bulk of the population, unable to comprehend the uncertainties implicit in a one-dimensional atmospheric model, and therefore needlessly alarmed by the concept (Maddox 1988). A second group of investigators, while accepting the basic premise of the TTAPS scenario, re-examined it with the intention of removing some of its shortcomings, mainly through the use of improved modelling techniques (Crutzen *et al.* 1984, Covey *et al.* 1984, Cess 1985). A third group broadened the scenario by examining its environmental and societal consequences (Ehrlich *et al.* 1983, Sagan 1983). The TTAPS report also encouraged broader studies of the global effects of large-scale nuclear war. These included an appraisal of the situation by such bodies as the Royal Society of Canada (RSC 1985), the US National Research Council (NRC 1985), and the Scientific Committee on Problems of the Environment (SCOPE) set up by the International Council of Scientific Unions (ICSU) (Harwell and Hutchinson 1985, Pittock *et al.* 1986).

Criticism of the TTAPS scenario

The TTAPS scenario elicited a strong response from the scientific community, but much of the initial criticism merely detailed the flaws in the study itself – many already acknowledged by Turco *et al.* – and provided little or no development of the problem of nuclear winter (Teller 1984). The assumptions underlying the estimated duration and intensity of smoke and dust clouds were strongly critized, and the magnitude of the postulated temperature decline was regarded as questionable (Maddox 1984, Singer 1984). Many of the criticisms were qualitative or speculative, although apparently based on well-established meteorological principles (e.g. Hainsworth 1985). Simple observation would suggest that if the heat of the sun could not penetrate the layers of smoke and dust then the stored heat emitted from the earth and oceans would be unable to escape either. Minimum temperatures are almost always higher on winter nights with cloud cover, than on nights when the skies are clear. Is there any indication that the smoke and dust from a nuclear explosion would act any differently from a natural cloud cover? Much would depend upon the characteristics of the particulate matter thrown into the atmosphere by the explosion or contributed by the accompanying fires. Carbon-rich soot is more effective at absorbing and scattering solar radiation than dust, for example. Thus, the greater the proportion of smoke in the atmosphere, the greater the potential cooling. Smoke particles with a diameter of about 1 μm absorb visible light well, but are less able to absorb energy in the infrared wavelengths. High levels of smoke therefore enhance cooling by allowing terrestrial infrared radiation to escape from the atmosphere (Bach 1986). In contrast, the water droplets in natural clouds trap and re-emit infrared radiation, helping to reduce the rate of cooling.

In the consideration of nuclear winter, the amount, distribution, and residence time of the nuclear smoke and dust veil in the

173

atmosphere are very contentious issues. Uncertainties involving smoke and dust exist at all levels in the development of the hypothesis. The nature of the nuclear exchange itself would influence the amounts of particulate matter injected into the atmosphere. A few well-distributed explosions would cause smoke and dust concentrations to remain patchy, allowing solar radiation to reach the earth's surface for some time after the original exchange. The resultant temperature variations would intensify local or mesoscale circulation patterns, helping to cleanse the atmosphere before the aerosols were evenly spread. Concentrations on military targets would also produce less smoke and dust than a full-scale war which included city targets. The smoke from urban areas, which contains high levels of elemental carbon produced by the burning of petroleum and plastic products, is much more effective in blocking radiation than smoke from natural grass and forest fires (Crutzen *et al.* 1984). The density of combustible material in urban areas is also more likely to produce high intensity fire-storms capable of convecting aerosols into the stratosphere (Teller 1984).

Turco *et al.* (1983) estimated an injection of 960 million tons of dust and 250 million tons of smoke into the atmosphere from their baseline case of a 5,000-megaton conflict. Teller (1984) implied that such estimates were too high, but quoted a range of 15 to 360 million tons of smoke. In a very detailed compilation of aerosol production following an exchange of close to 5,000 megatons, Crutzen *et al.* (1984) estimated smoke emissions of between 140 and 320 million tonnes.

The total impact of such emissions will depend not only upon the amounts involved, but also upon the rate at which they are removed from the atmosphere. The residence time of particulate matter in the atmosphere is influenced by the size and nature of the particles present. Large particles fall to the ground soon after they are emitted. Smaller particles collide and adhere to each other, forming larger particles which in turn fall out of the atmosphere. This process is called coagulation. It helps to clean the atmosphere, but, since larger particles are less able to scatter solar radiation, it also has an effect on the energy budget, even if the newly formed particles do not fall out immediately. One of the most efficient processes for removing smoke particles from the atmosphere is precipitation scavenging, caused when particles interact with water vapour, rain, and snow. The particles act as condensation nuclei or become attached to the water droplets and snowflakes to be carried out of the atmosphere in precipitation.

Precipitation scavenging is recognized as one of the processes by which the atmosphere begins to rid itself of the contamination produced by major conflagrations. It is known to accompany large forest fires, for example (Crutzen *et al.* 1984). The non-nuclear bombing raids on Hamburg and Dresden in the Second World War created firestorms which were followed by heavy precipitation. The rain at Dresden was described as 'sooty' and the rapid clearing of the smoke cloud caused by the fire is considered by Peczkic (1988) to be an indication of the efficiency of precipitation scavenging. 'Black rain' also fell at Hiroshima and Nagasaki following the nuclear bombing of these cities (Crutzen *et al.* 1984). That type of evidence suggests that precipitation scavenging would accompany any major nuclear exchange, and begin the process of clearing the atmosphere very soon after the initial explosion.

The TTAPS study has been strongly critized for underestimating the ability of the atmosphere to cleanse itself (Singer 1984, Teller 1984). Singer argued that most of the aerosols would be removed from the atmosphere long before they could be dispersed sufficiently to produce the radiation reduction necessary for nuclear winter. Teller pointed out that the weight of water injected into the atmosphere during a major nuclear war might be ten thousand times the weight of smoke. Such an abundance of water would ensure that almost all of the smoke would be washed out in a very short time, perhaps as little as a week. Teller also noted the important role of particulate matter in providing condensation nuclei which encourage precipitation, but there is some evidence that a super abundance of condensation nuclei may suppress precipitation by allowing the formation of many water droplets, most of which are too small to fall as rain (Bach 1986).

Singer's conclusions on precipitation scavenging were similar to those of Teller. He also concluded, however, that the abundant water vapour injected into the atmosphere would influence the earth's energy budget through its ability to absorb infrared radiation. This would contribute to the greenhouse effect, and counter, to some extent, the solar radiation reduction caused by the smoke and dust clouds.

The criticisms of Teller (1984), Singer (1984), and Maddox (1984) were based on the many uncertainties included in the TTAPS model of nuclear winter. In general, the critics concluded that Turco *et al.* (1983) over-estimated the amount of cooling likely to follow a major nuclear war, but their own estimates are no more verifiable than those in the TTAPS study. In its appraisal of the environmental impact of nuclear war, the Royal Society of Canada (1985) concluded that many of the specific criticisms had some substance, but overall they did not invalidate the concept of nuclear winter.

Modelling nuclear winter

Many of the criticisms of the TTAPS study can be traced to inadequacies in the one-dimensional (1-D) model of the atmosphere used in the original calculations. Subsequent research into nuclear winter has involved attempts to improve the hypothesis by developing better models of the processes involved.

Like all 1-D models, the TTAPS model treated the earth as a uniform surface with no geography and no seasons. It had the ability to estimate the impact of atmospheric aerosols on sunlight and infrared radiation absorption, which allowed temperatures to be calculated. It was inadequate to deal with the uneven surface-energy distribution associated with the differences in heat capacity between land and ocean, however. A major problem with the TTAPS model (shared with all 1-D systems) was its inability to deal with earth/atmosphere feedback loops. No account was taken of the impact of changing heat distribution on the dynamics of the atmosphere as the smoke and dust spread; it was assumed that circulation patterns would retain their pre-conflict form.

Crutzen *et al.* (1984) also used a 1-D model in their consideration of the climatological effects of post-nuclear fires. However, they included a more detailed examination of the sources and distribution of smoke during large nuclear fires than Turco *et al.* (1983). Their model incorporated a three-layer division of the atmosphere, based on smoke content, and included greater consideration of water and CO_2 levels. The simulation results generally supported the TTAPS study, but the model remained one-dimensional – with all the inherent drawbacks of the type.

One-dimensional models are useful for preliminary investigation of global scale radiative and convective processes at different levels in the atmosphere. However, they cannot deal with seasonal or regional scale features, and require so many assumptions that their ability to provide accurate predictions is limited. This has led to the development of two-dimensional (2-D) and three-dimensional (3-D) models.

Two-dimensional models add a meridianal component to the altitudinal component of the 1-D models. They can consider variations in dust and smoke along a vertical cross-section from pole to pole, for example. They also allow the redistribution of particulates in the atmosphere to be represented, although only in a relatively simple form (RSC 1985). MacCracken (1983) used a 2-D model to simulate the effects of smoke spreading from nuclear explosions. His model allowed him to include consideration of differences in heat capacity between land and ocean, but its ability

176

to deal with the evolving dynamics of the atmosphere as the smoke spread remained limited. The temperature reductions predicted by this model were generally less than those for the TTAPS model, but the cooling was expected to last longer. Other 2-D climatic models have been developed by the Canadian Atmospheric Environment Service and by the Lawrence Livermore National Laboratory in the United States, but they were superseded by 3-D models at an early stage in the study of nuclear winter (Evans *et al.* 1985).

Three-dimensional models provide full spatial analysis of the atmosphere. They involve the solution of the dynamic equations which represent large-scale atmospheric motion, and can include the important feedback mechanisms missing from 1-D and 2-D models. Climate models of this type are know as general circulation models (GCMs). Although their treatment of atmospheric dynamics is quite sophisticated, other elements – such as radiation – are often less finely developed than in 1-D models.

One of the first of the 3-D models used to examine the problem of nuclear winter was developed as a GCM by the National Centre for Atmospheric Research (NCAR). Covey *et al.* (1984) used it to deal with the impact of smoke injections into the atmosphere during a nuclear war. It included a greater number of variables and finer geographic resolution than the TTAPS model. Ozone, CO_2, and water vapour were incorporated in the model, and allowance was made for the presence of cloud in the lower troposphere. It could also cope with information at nine different levels in the troposphere and stratosphere, but retained some of the basic uncertainties of earlier models. For example, initial differences in the heat capacity of land and water were acknowledged, but no attempt was made to deal with changes in their relative values as the scenario progressed. The model dealt with smoke only. It did not consider the dust likely to be lofted into the atmosphere during a major nuclear exchange.

When the model was run, the twin Hadley cells normally lying on either side of the equator were replaced by a single cell rising in the northern hemisphere and descending in the southern hemisphere. This was brought about by the intense heating of the upper atmosphere as the smoke cloud in mid-latitudes absorbed solar radiation. Such a radical change in the circulation of the atmosphere would facilitate the spread of smoke south of the equator, allowing the effects of surface cooling to spread also. According to Thompson (1984), this may be a function of the 'smoke-only' format adopted for the model. He has suggested that the presence of dust mixed with the smoke would reduce the absorption of solar radiation in the stratosphere and upper troposphere. This would dampen the temperature increase in the upper atmosphere, and effectively prevent the development of the single cell.

It is evident that the replacement of a 1-D model with a more complex 3-D model is no guarantee that the new predictions will be an improvement on the earlier ones. In reality, the GCM used by Covey *et al.* (1984) generally supported the TTAPS results, although the predicted cooling was less because of the ameliorating effects of stored heat in the oceans. One important conclusion of the GCM study was the recognition of the dynamic nature of the problem. The smoke produced during a nuclear exchange could not be expected to remain a passive constituent of the atmosphere for long. Initially, its distribution would be controlled by the existing atmospheric circulation, but the changing temperature structure of the atmosphere – caused by the ability of the smoke to absorb solar radiation – would eventually alter that circulation. The smoke would then be dispersed by the new smoke-induced circulation. GCMs incorporating fully interactive aerosol schemes are now available to deal with such complexities. They are the most sophisticated models yet developed to simulate the climatological effects of nuclear war.

State-of-the-art simulations have been performed using an NCAR interactive GCM which can deal with the atmosphere from the surface to an altitude of 30 km, and which resolves geographic features at about 5° latitude by 7° longitude. The model includes transportation of aerosols by winds, removal of particles from the atmosphere by rainfall and other processes, detailed calculations of sunlight transmission, and infrared greenhouse effects for both dust and smoke (Thompson 1985). Simulations using this system indicate that, after a 6,500-megaton exchange, the planetary cooling would be much less than that postulated in the TTAPS scenario. The difference may be explained by a greater allowance for the heat capacity of the oceans, more rapid removal of atmospheric smoke, and the inclusion of the greenhouse effect – all of which were less than adequately considered in the original study. Temperatures returned to normal as little as 30 days after the initial exchange in some of the NCAR simulations, and it has been suggested that the climatic perturbations associated with nuclear war might be referred to more appropriately as 'nuclear autumn' (or 'nuclear fall') rather than nuclear winter (Thompson and Schneider 1986) (see Figure 8.2).

Despite the seeming sophistication of the latest GCMs, they remain crude compared with the actual workings of the atmosphere. No model can be better than the understanding of the

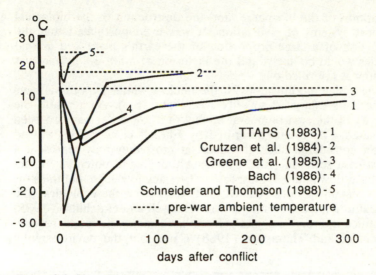

Figure 8.2 Changing estimates of the intensity and duration of nuclear winter

atmosphere on which it is based, nor can the results obtained be better than the data fed into it. The simulations used to predict the climatological impact of nuclear warfare are delinquent on both counts.

The perfect model will remain out of reach until the processes and mechanisms which drive the earth/atmosphere system are better understood. Even with the perfect model, however, the many uncertainties in data input – including the number, location, and yield of the nuclear explosions, or the relative proportions of smoke and dust, for example – would limit the accuracy of the simulations.

The environmental and societal consequences of nuclear war

The potential biological impact of nuclear winter was noted in the TTAPS scenario, but was not elaborated. At about the same time as the atmospheric scientists were developing the hypothesis, however, a group of life scientists was also examining the long-term, world-wide, biological consequences of nuclear war. Some forty of the world's leading biologists met to discuss the topic in April 1983, and the results of their deliberations were presented later that year (Ehrlich *et al.* 1983). It was their considered opinion that the net effect of a large-scale nuclear war would be the global disruption of the biosphere, and the destruction of the biological support systems of civilization. It was even postulated that the extinction of a large proportion of the earth's plant and animal species would be likely, and the inclusion of *homo sapiens* in that group was not ruled out.

Successive modifications of the nuclear winter hypothesis have reduced its estimated severity (see Figure 8.3). As a result, the severity of the environmental impact of nuclear war has also been mitigated. The events described by Ehrlich *et al.* (1983) are no longer considered likely to occur; at most they may represent a worst-case scenario (RSC 1985). Although the potential impact has been downgraded, most researchers acknowledge that the situation remains serious. The SCOPE Report on the ecological and agricultural impact of nuclear war (Harwell and Hutchinson 1985) identifies indirect biological effects as a major threat to society. Schneider and Thompson (1988) claim that the environmental

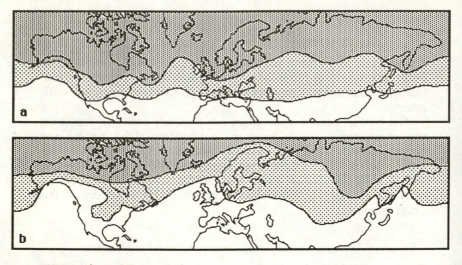

< 0°C   0-10°C

Figure 8.3 Different estimates of temperature distribution following a major summer nuclear conflict: (a) represents the classic nuclear winter scenario; (b) represents the less serious scenario calculated using improved modelling techniques

Sources: (a) compiled from data in Turco *et al.* (1983), Aleksandrov (1985); (b) after Schneider and Thompson (1988)

effects in their less drastic nuclear autumn scenario are still capable of producing catastrophic consequences, and strong critics of the original TTAPS theory, such as Teller (1984), acknowledge that even if the temperature fell by only one tenth that postulated by Turco *et al.* (1983), it could still lead to widespread agricultural failure.

Life on earth is dependent upon an adequate supply of solar energy. Sunlight processed through green plants during photosynthesis provides energy for animals and humans. Any large-scale reduction of solar radiation or destruction of plant life will ultimately reduce the ability of the earth to sustain life. In a nuclear war, blast and fire would destroy vegetation cover over large areas. Surviving plants would be subject to low light conditions and falling temperatures. Most would cease to grow, and many would die as photosynthesis fell below the level required to meet the plants' metabolic needs (Grover 1985). The rapid cooling would also contribute to the problem since photosynthesis is less efficient at lower temperatures (see Figure 8.4).

In addition to their impact on photosynthesis, low temperatures would damage plants directly. The impact would vary with the season. Many plants – such as those in the boreal forests of Canada, northern Europe, and the USSR – can survive low temperatures when they are dormant, but even very small temperature changes will cause severe damage if they occur during the growth cycle (Ehrlich *et al.* 1983). Thus, the greatest damage to vegetation would occur if the nuclear exchange took place during the growing season. Abrupt freezing in mid to high latitudes would cause complete destruction of many species normally resistant to winter cold. The survivors would have their productivity much reduced, and recovery would be retarded by the persistence of freezing temperatures for several months after the original attacks.

A nuclear war during the northern hemisphere's winter would be less damaging to vegetation in mid to high latitudes. The effects might be carried over into the following spring, however, when the continued cooling would retard the start of the growing season (Turco and Ackerman 1985). It is likely that vegetation in tropical regions would suffer significant damage whatever the season. Tropical plant systems flourish under conditions which include mild, stable temperatures. They are susceptible to moderate declines in temperature, and do not develop the resistance to cold which helps temperate plants to survive (Greene *et al.* 1985). Ehrlich *et al.* (1983) have suggested that forests in tropical regions might largely disappear if low light levels and low temperatures were to become widespread.

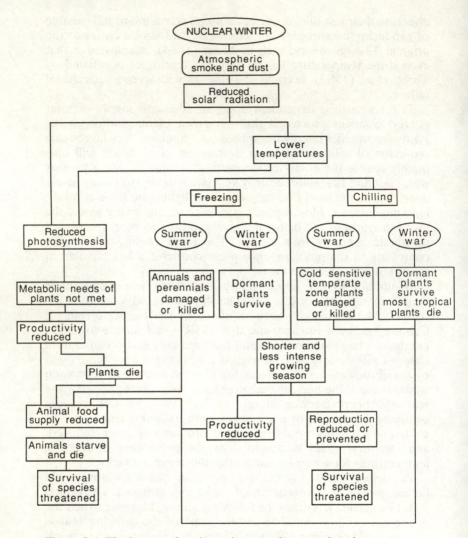

Figure 8.4 The impact of nuclear winter on the natural environment

Marine ecosystems would be more capable of dealing with temperature extremes, because of the moderating influence of the oceans. Trophic webs would be seriously disrupted by the reduced productivity of phytoplankton during the low-light period, however, and the increased storminess in coastal areas would cause damage to shallow-water ecosystems (RSC 1985). Overall, it is considered that aquatic ecosystems have the potential for a more rapid recovery than terrestrial systems. In the latter, even a decade after the event, structures and processes would remain unstable, and the systems unbalanced (Ehrlich *et al.* 1983).

The damage to vegetation in natural ecosystems would be paralleled by damage to cultivated plant species. Tropical food crops such as rice, bananas, and maize can be damaged by temperatures falling to 7–10 °C for as little as a few days, and even that moderate chilling would be sufficient to cause crop failure (Greene *et al.* 1985). In temperate latitudes, wheat and corn would suffer much the same fate. Nuclear war during the growing season would wipe out the grain crop in the Canadian Prairies and the northern Great Plains of the United States (Grover 1985). Damage to agricultural crops on such a large scale would have a major impact on world food supply. Famine would occur in those areas directly involved in the nuclear exchange, but even areas far removed from the conflict would ultimately suffer.

Many tropical Third World countries already experience food shortages, which would be aggravated by the effects of chilling. The food surpluses from North America and Europe, upon which they have come to depend to alleviate such shortages, would no longer be available. Even if largely unaffected by low light or chilling, indigenous agriculture in the Third World would experience a reduction in productivity as the supply of fertilizer usually obtained from the developed nations was cut off. The cumulative effects of such developments would be disastrous, and it is not surprising that the SCOPE Report (Harwell and Hutchinson 1985) has identified mass starvation as the major consequence of nuclear war for humankind.

In addition to the low light levels, sub-freezing temperatures, and violent storms, the human survivors of the initial conflict would face continued radioactive fallout, high levels of toxic air pollution, and enhanced levels of ultraviolet radiation. Combined with lack of food and drinking water, psychological stress, and the disruption of support systems such as transportation, communications, and medical care, this would probably ensure that fatality rates remained high even years after the conflict.

The current status of nuclear winter (1988)

Nuclear winter has been the focus of a remarkable amount of attention since it was first postulated by Richard Turco and his collaborators in 1983. The world's leading atmospheric scientists have examined its physical base in detail, and life scientists have used these findings to estimate the impact of nuclear war on the environment. The problem has been studied by the world's major government-funded research agencies, such as the Academy of Sciences in the USSR and the National Center for Atmospheric Research in the United States (Schneider and Thompson 1988). Funding from the United States Department of Energy and the Defence Nuclear Agency has supported investigations of the topic at such prestigious institutions as the Los Alamos and Lawrence Livermore National Laboratories (Teller 1984, Schneider 1988). The net result of this intense activity has been the convening of several conferences, and the publication of close to 100 academic papers on the subject between 1983 and 1988 (Bach 1986, Schneider and Thompson 1988). Numerous newspaper and magazine articles have also appeared.

The most obvious effect of all of this has been the modification of the level of cooling to be expected. The TTAPS scenario included an estimated summertime cooling of as much as 35 °C in mid-latitudes, but subsequent studies, employing more sophisticated models than that used by Turco *et al.* (1983), have reduced that to between 5 °C and 15 °C (see Figure 8.2). Under such conditions, summer temperatures in mid-continental areas might not fall below freezing (Schneider and Thompson 1988) (see Figure 8.3). However, several models have identified the possibility of transient quick freezes lasting for several days even in midsummer and associated with the effects of local weather conditions (Covey *et al.* 1984, Schneider 1987). Cooling by even the average amounts would be sufficient to cause freezing during the other seasons of the year.

These short-term effects taking place within the first month after the conflict have been identified as 'acute' effects, in contrast to the longer term 'chronic' effects (Pittock *et al.* 1986). In most cases the acute effects have been considered to have the greatest climatological impact because they include the initial temperature decline. As more work is done on the longer-term climatological effects of nuclear war, however, there is evidence that the chronic effects could be more serious than once thought (Schneider 1988). Chronic effects are produced by the smoke remaining in the atmosphere after several weeks of precipitation scavenging. Most of the remnant smoke would be found in the stratosphere, and the amounts involved would vary according to the magnitude and timing of the nuclear conflict. A relatively small-scale winter war would send little smoke into the stratosphere. In contrast, a full-scale summer conflict, with many urban targets, would inject large amounts of smoke beyond the tropopause, and several weeks after the end of the war as much as 50 per cent of the smoke produced might remain behind in the upper atmosphere (Schneider and Thompson 1988). The persistence of this stratospheric smoke for several years might reduce the amount of solar radiation reaching the earth's surface sufficiently to create and maintain an average global cooling of several degrees. Thompson and Schneider (1986) have suggested that this would ensure the increased probability of late spring or early fall frosts. The longer the smoke remains, the more likely it is to become involved in the larger scale of dynamics of the earth/atmosphere system – through feedback mechanisms, for example. Robock (1984) has examined the role of snow and ice feedbacks in prolonging cooling after a nuclear war, and although many uncertainties remain, the possibility that residual smoke could initiate long-term climatic change cannot be ruled out.

Another development in the study of the climatic effects of nuclear war has been the consideration of precipitation. The TTAPS study said little about precipitation beyond its ability to wash particulate matter out of the atmosphere. Recent simulations have considered the effects of cooling on large-scale precipitation distribution. In general, the stabilization of the earth/atmosphere system produced by the blocking of solar radiation by smoke in the upper atmosphere would be expected to reduce the vigour of the atmospheric mixing, and therefore lower the global average precipitation rate (Bach 1986). The reduction would not be evenly distributed, however. In the first 30 days after the conflict, average rainfall reductions of 20 – 50 per cent might occur in mid-latitudes in the northern hemisphere, with the greater reductions occurring over land (Schneider and Thompson 1988). The rising temperatures in the upper troposphere, resulting from the increased absorption of solar radiation, would retard condensation – causing the normally rainy tropics to experience reductions also (Thompson 1985). It is possible that following a nuclear war the cooling of the continents would disrupt the monsoon circulation in Asia and Africa. A summer war could reverse the monsoon circulation, leading to unseasonably dry conditions during the normal rainy season (Bach 1986). The chronic effects of nuclear war might even allow that situation to persist, causing a reduction in monsoonal rains for several years (Pittock *et al.* 1986).

Many uncertainties remain, despite the progress made in developing accurate simulations. Both the extent and intensity of a possible nuclear war are unknown or predictable only over a very broad range. As a result, inputs basic to the working of the models can only be approximate. It is clear, for example, that the role of smoke is crucial in any assessment of the consequences of a nuclear conflict. It is almost impossible, however, to predict the amount of smoke that will be produced without some knowledge of the targets chosen and the number of warheads exploded. Nor is it possible to predict the proportion of smoke that will be injected into the upper atmosphere. Once the smoke is in the atmosphere the rate at which it is removed will also be crucial, but that too remains uncertain (Bach 1986). Under such circumstances, even perfect models would be unable to produce completely reliable predictions.

Most models are not perfect. They continue to be inadequate in many respects. Some of the GCMs do not include such potentially important elements as low-level cloudiness and radiation fog formation (Bach 1986). Others include complex processes – energy flow close to the surface, for example – but in such a simplified form as to affect the quality and accuracy of the simulations (Schneider and Thompson 1988). Even with all of these uncertainties, however, climate models provide the only realistic method of simulating the climatological impact of nuclear war. Inaccurate they may be, but they are certainly preferable to the real thing.

The changes which have taken place in the interpretation of the climatological impact of nuclear war have been carried through into the consideration of its biological and societal impact. When Ehrlich *et al.* (1983) first examined the long-term biological consequences of nuclear war they came to the conclusion that the environmental effects of nuclear winter would be at least as serious as the effects of blast, fire, and radiation. That conclusion was echoed in the SCOPE Report which suggested that indirect effects such as mass starvation would be of greater consequence to human beings than the direct effects of the nuclear explosions (Harwell and Hutchinson 1985). Among the combatant nations, however, it seems unlikely that the indirect effects could be any more devastating than the original explosions (Schneider and Thompson 1988), but outside the main battle area the indirect effects would dominate (Schneider 1988). Indirect environmental effects resulting from climatic change would work in conjunction with socio-economic effects – such as the disruption of trade in energy, food, fertilizers, and medicine – to threaten large numbers of people in Asia, Africa, and South America. As the severity of nuclear winter has been downgraded, so has its impact on society. The extinction of the human species is no longer considered likely, for example, although the potential for a disastrous disruption of human activities remains. There is also enough evidence from the models to suggest that, even in the less destructive scenarios, existing problems of acid rain, ozone depletion, and the greenhouse effect pale before its capacity to destroy – perhaps permanently – the earth's environment.

Summary

The concept of nuclear winter grew out of the TTAPS study of the global consequences of multiple nuclear explosions. It postulated that a major nuclear conflict would be followed by a very rapid cooling of the earth, sufficient to cause freezing temperatures in some areas even in mid-summer. This would be caused by the very effective blocking of incoming solar radiation by the smoke and dust thrown into the atmosphere by the explosions. Because of its detrimental effect on the environment it was suggested that nuclear winter might have a greater impact on the earth and its inhabitants than the original nuclear explosions.

Of necessity, the development of the concept was entirely theoretical, based on results from a series of physical models. Much of the activity among researchers studying nuclear winter has involved the improvement of models and the refinement of modelling techniques. Although none of the models is perfect, the most sophisticated provide sufficiently similar results that a general concensus has emerged. The severity of nuclear winter has been downgraded and its environmental impact is no longer considered to be the unmitigated disaster it once was. It has not been downgraded so much that it can be ignored, however. Even the most optimistic simulations suggest that global cooling, along with the disruption of other climatic elements, would have serious environmental and socio-economic consequences.

Suggestions for further study

1. Obtain a list of mean daily temperature normals for your area from the nearest weather office. Starting with the mean daily temperature for 1 January, and using the same axes and scales as in Figure 8.2, draw the normal temperature curve represented by your data. Superimpose on the graph a curve based on the temperature depression and recovery times postulated by TTAPS (1983) to indicate the greatest climatic impact of a nuclear exchange. Add a third curve to represent the estimates of Schneider and Thompson (1988). Repeat the process for 1 April, 1 July, and 1 October using the mean daily temperature on each of these days as a starting point.

Use the four graphs to assess the impact of nuclear winter in your area. Compare the TTAPS scenario with the less serious situation postulated by Schneider and Thompson. Consider such factors as changes in the timing of freeze-up and break-up, the occurrence of early and late frosts, and the length of the growing season. What are the limitations of this type of assessment?

2. Write a short account of the role of the nuclear winter hypothesis in the nuclear disarmament debate. Include the following in your considerations: the total tonnage of nuclear explosive devices currently available; the effects of the US/USSR nuclear weapons treaties; the impact of nuclear winter on the concept of a winnable nuclear war; the criticism that the nuclear winter hypothesis was needlessly alarmist; and the suggestion that the downgrading of nuclear winter to nuclear fall (autumn) ensures that the climatic effects of nuclear war will be minor compared to the effects of blast, fire, and radiation.

Chapter nine

Present problems and future prospects

Public interest in the global environmental issues described in the preceding chapters has waxed and waned over the past decade. At present, acid rain, ozone depletion, and the greenhouse effect elicit a high level of concern, whereas atmospheric turbidity, drought and desertification, and nuclear winter have a much lower profile than they once had. This reflects current perceptions of the seriousness of particular problems. Perceptions can change, however. Since few members of the general public are in a position to read the original scientific reports which address the issues, they must depend upon an intermediary to satisfy their interest. In modern society this interpretive role has been filled by the media, and public perception of the issues is formed to a large extent by their interpretation of research results. Without them, the general level of understanding of the problems would be much lower than it is, but, as a group, they have also been accused of sensationalizing and misinterpreting the facts supplied by the scientific community. There can be no doubt that some of the accusations are valid, but scientists themselves may be partly to blame for allowing conclusions to be presented as firm, before all of the facts are in. Such was the case with nuclear winter, and also with the recent discovery of the hole in the ozone layer above the Antarctic. Given the scale and complexity of current environmental issues, problems of interpretation and dissemination are inevitable. They must not be allowed to divert attention from the main task, however, which is the search for solutions to the major issues.

Current state of the issues

One of the first environmental issues to be considered in a global context was the rising level of atmospheric turbidity, which was the centre of concern in the mid-1970s. It linked air pollution with the cooling of the earth. Cooling had been taking place since the

1940s, and some writers saw the world descending into a new Ice Age. A decade later, it was clear that the cooling had reversed, and atmospheric turbidity began to receive less attention. Evidence also began to appear which indicated that increased atmospheric turbidity might actually contribute to atomospheric warming. Currently, it raises little concern among the general public. The conditions which caused the turbidity to increase remain, however, and scientists interested in the impact of human activities on the atmosphere continue to study it.

Interest in nuclear winter has also waned from a peak reached between 1983 and 1985. Intensive investigation of the issue, from 1983 onwards, produced increasing evidence that the climatic impact of nuclear war had been overestimated in the original study. The downgrading of the estimates in the scientific and academic community was inevitably accompanied by a general decline in the level of public interest in the topic. Recent agreements between the US and the USSR on nuclear-weapons reduction have created the perception that nuclear war is now less likely, and this too has contributed to the decline. As a result of such changes, nuclear winter is now regarded as irrelevant by some, but in reality little has changed. Scenarios formulated by the computer-model builders who remain interested in the nuclear hypothesis are much less daunting than the original, but the projected climatic deterioration remains sufficiently serious to cause major problems for the nations involved in any conflict, and for their neighbours. In addition, the nuclear-weapons reduction currently taking place involves only a limited number of devices. The nuclear nations retain enough bombs and warheads to create the conditions required for nuclear winter several times over.

Drought, famine, and desertification are related problems of long standing in many parts of the world. The disastrous droughts in the Sahel in the 1960s and 1970s, for example, were only the most recent in a series which can be traced back several centuries. The earlier droughts, and their accompanying famines, passed mostly unnoticed outside the areas immediately affected. In contrast, modern droughts have been characterized by a high level of concern, particularly in the developed nations of the northern hemisphere. Concern is often media-driven, rising rapidly, but falling just as quickly when the drought breaks or the initial benefits of food and medical aid become apparent. When the rains returned to the Sahel in the late 1970s, interest in the problems of the area declined, although the improvements were little more than minimal. Similarly, the concern raised by television reports of drought and famine in Ethiopia in the mid-1980s peaked at a very high level in 1985 with the Live Aid concerts, only to decline again within the year. Such fluctuations give a false impression of the problem. Drought and famine are endemic in many parts of the world, and do not go away when the interest of the developed nations declines. In some areas the return of the rains does bring periodic relief, and the land produces a crop. Where the relief is infrequent or short-lived, the desert advances inexorably into previously habitable land. This is desertification in its most elemental form. Accelerated by human interference, it has become the most serious environmental problem facing some of the countries of the earth's arid zones. Climatologists, agronomists, foresters, and the scientists of the United Nations Food and Agricultural Organization have been wrestling with it for nearly 40 years, yet even now it receives less public recognition than the drought and famine with which it is associated.

Public interest in drought, famine, and desertification will continue to fluctuate. Heightened concern, followed by increased financial and nutritional aid, may help to alleviate some of the immediate effects of the problems, but it is the steady, less volatile interest of the scientific community which has the potential to bring about longer-term relief.

The problems of acid rain, ozone depletion, and the enhanced greenhouse effect currently enjoy a much higher profile than the other environmental issues. All three are the focus of major research efforts. Millions of dollars have been invested in an attempt to identify the causes and effects of the problems, and to suggest possible solutions. Interest in the topics has been developed and maintained through a continuing series of high-level international conferences.

Since 1979, for example, there have been at least four major international meetings on acid rain. These led to the formation of the so-called '30 per cent club' which requires its members to reduce trans-boundary emissions of SO_2 by 30 per cent (of 1980 levels) before 1993. National and regional conferences have complemented the international meetings by dealing with specific aspects of the problem.

Ozone depletion has been treated in much the same way. The Vienna Convention on the Protection of the Ozone Layer, which emerged from a 1985 conference, was followed in 1987 by the Montreal Protocol. Signatories to the Protocol agreed to reduce production of CFCs by 50 per cent (of 1986 values) by the year 2000. The concluding statement of the World Conference on the Changing Atmosphere – held in Toronto, Canada, in mid-1988 – also included reference to CFCs, and called for them to be phased out by the year 2000. Later in 1988, delegates at the World Conference on Climate and Development in Hamburg, West Germany, recommended a global ban on the production and use of CFCs by 1995. In March 1989, the members of the European Economic Community agreed to eliminate the production and use of ozone-destroying chemicals by the end of the century. The US government endorsed the effort, but stressed the importance of finding safe substitutes for CFCs. At about the same time, major government-sponsored conferences in London and Paris provided international support for a world-wide ban on CFCs and other environmentally harmful chemicals.

Both the Toronto and Hamburg conferences also considered the impact of an enhanced greenhouse effect on the world. The United Nations Environment Programme (UNEP) and the World Meteorological Organization (WMO) created a 35-member Inter-governmental Panel on Climate Change (IPCC) in late 1988 to evaluate global climatic trends, with an emphasis on global warming, and this was followed in April 1989 by a Seminar on Global Climate sponsored by Mrs Margaret Thatcher, the British Prime Minister. These are the most recent in a series of conferences and hearings which have examined this topic. Some have dealt with it at the global scale, others have considered the local impact, but the net effect has been the accumulation of a considerable body of knowledge on all aspects of the greenhouse effect. Much of the material is complex – written in the scientific jargon of research reports – but, perhaps more than any of the other issues, the greenhouse effect has been treated by the media in such a way as to stimulate public interest in the problem. Government organizations have also published the evidence from the research reports in a simplified form for media use and public consumption. The Climate Change Digests of the Canadian Atmospheric Environment Service are a good example of this approach. In addition, private, non-profit organizations – such as the Climate Institute in the United States – have been formed to advance public understanding of the global warming produced by the greenhouse effect.

The success of these endeavours is difficult to measure, as yet, but the approach is becoming more common. It reflects the general consensus in the scientific community that solutions to current large-scale environmental problems can be implemented successfully only if they have a high level of public support, and that such support is most likely to come from a public kept well informed about the nature and extent of the problems.

Solutions

The urgent need to provide solutions to current global environmental problems is widely accepted, but there is a minority viewpoint which suggests that society could cope by making appropriate adjustments to lifestyles. Although this approach may be necessary until the full effects of the solutions can work their way through the system, it can only be a temporary measure. In some cases adjustment is difficult or even impossible. How does society easily cope with the increased levels of ultraviolet radiation caused by ozone depletion, for example? It might be possible to cope successfully with other issues in the short term, but ultimately the cumulative effects of the changes would surpass the ability of society to adjust, and solutions would have to be found.

When considered qualitatively, in the academic isolation of the lecture hall or the comfort of an armchair, it appears that all of the current global environmental problems can be solved. They share the same overall cause, human interference in the environment, and specific causes are common to several of the issues. Acid rain, increased atmospheric turbidity, the intensification of the greenhouse effect, and ozone depletion could all be reduced with the exercise of greater control over anthropogenic emission of dust, smoke, and gases. Energy consumption is also an element common to many of these problems. The reassessment of energy sources, leading eventually to a reduction in fossil-fuel use, is seen by many to provide the most likely solution to the problems of atmospheric turbidity, acid rain, and the intensification of the greenhouse effect. It is particularly attractive because it is a broad-spectrum solution, which does not require separate technology to be developed for each issue. Those problems which could not be solved completely might have their effects mitigated. It is unlikely, for example, that mankind will ever be able to prevent or control drought, but, through the management of human activities in areas prone to drought, problems of famine and desertification might be alleviated. Nuclear winter, potentially the most devastating threat, yet in many ways the easiest to prevent, could be removed by an extension of the bilateral agreements on nuclear disarmament already being implemented by the US and USSR.

The technology exists to accomplish all of these, but the gap between theory and practice is immense, perhaps insurmountable. Much of the difficulty arises out of the socio-economic and political consequences of such changes, which are perceived by some to be even more detrimental to society than the continued existence of the problem. In short, the disease is considered less damaging than the cure. For example, a substantive diminution of acid rain could be accomplished by restricting the use of sulphur-rich, bituminous coal, but it would have the effect of placing in jeopardy the economic viability of the areas producing that commodity. The economic, social, and political impact of the closure of dozens of mines, accompanied by thousands of redundancies, might be seen as much more serious than the death of even several hundred lakes. Similarly, the problem of drought and famine might be approached by letting nature take its course, as was the norm in the not too distant past. It can be argued that the present system of providing food aid and drilling wells in areas suffering from famine and drought is the easiest way to ensure that the problems will continue well into the future. Cutting back on aid, or removing it completely, would lead to large-scale starvation and death, but it would also reduce stress on the environment, and allow the restoration of some form of equilibrium to the system. Current moral and ethical attitudes would presumably prevent the adoption of such a policy, and even the suggestion that it be considered would have far-reaching political implications.

Solutions to the global environmental problems presently facing society must be economically, socially, and politically acceptable, but they must go further than that. Current global problems are multifaceted, therefore the solutions must be multifaceted. They must consider societal and environmental consequences equally, rather than emphasizing the former, as is commonly done at present.

The environment suffers because of the time-scales followed by modern society. Politicians, for example, tend to deal in short-term causes, effects, and solutions, living as they do from election to election. Many environmental problems do not fit readily into such a framework. Rapid, sometimes catastrophic, change is an element in the environment, but, more often than not, change is accomplished through the cumulative effects of relatively minor variations over a long period of time. As a result, potentially important changes may not be recognized, or, if they are, they are ignored because of their seeming insignificance. Even when attempts are made to deal with such changes, the results may become apparent only after a considerable period of time. In the case of ozone depletion, for example, it would take as much as a decade following the complete abolition of CFCs before ozone levels would return to their normal range. In politics, where immediate and obvious solutions tend to be the order of the day, such a time lag is often considered politically unacceptable, and no action is taken.

Despite this, concern for the societal impact of environmental change has been growing among politicians and government bodies. They have made funds available for the investigation of global problems, and encouraged the dissemination of the research results through conferences and publications. The next stage in the process must be the implementation of the recommendations contained in the research reports. That is proving difficult, however, since it requires a long-term planning strategy, and modern planning policy is designed to provide solutions to short-term problems. Environmental impact procedures, for example, seek to integrate the environmental and socio-economic considerations arising from the development of a specific project and to mitigate their effects at the outset. It is assumed that the decisions made at that time will minimize environmental impact throughout the life of the project, but there is already evidence that environmental change is accelerating so rapidly that this approach is now inappropriate. If they are to be solved, current environmental problems require long-term planning which will extend several decades into the future and which will be responsive to change. Planners and policy-makers have not yet adjusted to that requirement. For example, reforestation is going ahead on the assumption that current climatic conditions will prevail during the life-span of the trees, yet in the next 50 years the intensification of the greenhouse effect is likely to cause the climate to change in those areas being planted. Irrigation projects and hydro-electric schemes costing millions of dollars are being developed with no thought for the impact of current global problems on precipitation patterns several decades from now. Coastal and waterfront property is being developed as if the rise in sea-level, projected to accompany global warming, is of no consequence. Few of the long-range plans necessary to deal with the problems have been put in place, and there have been few important governmental or industrial decisions which have paid more than lip-service to the recommendations of the research scientists.

The implementation of measures to alleviate the effects of global environmental disruption is further complicated by the scale of the problems. Most will require international co-operation if they are to be controlled successfully. The major conferences which have addressed such issues as acid rain, ozone depletion, and the greenhouse effect have been international in scope and have included agreements in principle on the measures required to reduce their impact. Such agreements are important, but they provide no guarantee that the situation will improve. Since they require ratification by individual nations, delays in their implementation are common. One year after the signing of the Montreal

Protocol on the depletion of the ozone layer, only seven of the original thirty-seven signatories had ratified the treaty. Even when there is complete ratification, the problem of enforcement remains, and the possibility always exists that the worst offenders may have refused to become involved. For example, Britain and the United States initially declined membership in the '30 per cent club' when it was formed to combat rising levels of acid emissions. Both are major contributors to acid rain, and many environmentalists feel that their lack of co-operation could be disastrous. Britain has subsequently revealed plans to reduce sulphur dioxide emissions by 60 per cent, but the United States remains slow to react. In North America, the fight against acid rain might well be lost if Canada and the United States do not come to some bilateral agreement soon. The Canadian government has already instituted measures which will lead to the reduction of acid emissions, but 50 per cent of the acid precipitation falling in Canada originates in the United States. Therefore, unilateral action can achieve only limited results. Despite signing an acid-rain control agreement with Mexico in 1987, the United States remains perversely slow in dealing with the problem along its northern border.

Similar problems arise with drought, famine, and desertification. These were originally local issues in the Third World but became global when the developed nations began to provide relief from drought and famine, and sought to combat desertification. Some success has been achieved against drought and famine, usually by employing the developed world's advanced technology and long-established supply and transport systems. Desertification remains rampant in many areas. Steps must be taken to prevent further environmental damage and to rehabilitate areas already damaged. Since it is independent of national boundaries, however, attempts to halt the spread of the desert in one area may be negated if nothing is done in an adjacent area. Success is possible only with international co-operation, and economic or political pressure may have to be applied to achieve that. Even if co-operation is complete, however, there is still no guarantee that the problem will be solved. Much will depend upon economic conditions in the developed nations for they will be required to provide much of the necessary financial aid. Any downturn in the world economy could put their contribution in jeopardy, and threaten the success of the fight against desertification.

Although the need for global co-operation to combat global environmental problems is widely recognized, it seems likely that the vagaries of international politics and economics will continue to frustrate attempts to implement solutions.

Further study: necessary or not?

Current global environmental problems are remarkably complex, and, despite an increasingly intensive research effort, they are even now not completely understood. That situation must change if solutions are to be found. Almost all individuals and organizations studying the problems have indicated that further study is necessary. In the past, that was often seen as a delaying tactic in that it was often easier to suggest further study than to make a positive attack on a problem. Attempts to solve the acid rain problem in North America were thwarted by these very tactics for many years.

Such arguments are no longer possible now that sufficient data are available, yet there remains a very real need for further study of many aspects of the earth/atmophere system. The roles of the various atmospheric processes require particular attention since they are intimately involved in all of the major problems currently confronting the environment. Traditionally, the study of the atmosphere was based on the collection and analysis of observational data. That approach is time-consuming and costly; it is also of limited overall accuracy because of gaps in the meteorological network, particularly in high latitudes and over the oceans. Modern attempts at exploring the workings of the atmosphere are almost exclusively dependent upon computer models, which range in complexity from simple 1-D formats providing information on one element in the system to highly sophisticated models employing as many as 100 variables, and including consideration of the oceans as well as the atmosphere. Studies of such topics as nuclear winter, the greenhouse effect, and ozone depletion have benefited greatly from the use of computer-modelling techniques. The models are becoming increasingly complex and comprehensive, but a high level of sophistication is no guarantee of perfection. Even state-of-the-art, 3-D general circulation models include some degree of simplification, and certain variables – cloudiness, for example – are very difficult to deal with whatever the level of model employed. In short, there is as yet no model capable of simulating exactly the conditions and processes in the real atmosphere. That should not be used as an excuse to do nothing, however. It might be tempting to wait for the perfect model, but that may prove impossible to develop. Existing models do have flaws, but they can be used provided their limitations are understood, and despite their inherent problems there is probably no better way of studying environmental problems involving the atmosphere.

Models have provided forecasts of the future extent of global environmental problems, and it is upon these forecasts that the strategies for dealing with the issues will be based. They concentrate on the impact of human interference on the earth/ atmosphere system, but generally ignore the possibility that natural variations in the system will instigate change. Most assume that the natural elements in the system will remain benign. Evidence from the earth's past suggests that this is an assumption which cannot be made. It is important that every attempt be made to identify and understand natural variations in the earth/atmosphere system, both at present and in the past. Only then will it be possible to make realistic projections into the future. That task is a difficult one. At present, weather-forecasting accuracy falls off sharply with anything beyond a 5-day forecast. How much success can be expected from a 50- to 100-year forecast? Perhaps very little, but if the form and function of the environment is to be retained, and even improved, the attempt must be made and made soon.

Bibliography

Aleksandrov, V.V. (1985) 'Global shield causes nuclear winter', in Royal Society of Canada, *Nuclear Winter and Associated Effects*, Ottawa: The Royal Society of Canada.

Almer, B., Dickson, W., Ekstrom, C., and Hornstrom, E. (1974) 'Effects of acidification in Swedish lakes', *Ambio* 3: 30–6.

Anon. (1983) 'One-third of German trees hit by acid rain', *New Scientist* 100: 250.

Anon. (1984) 'Balancing the risks: what do we know?' *Weatherwise* 37: 240–9.

Anon. (1988) 'CFC producers plan phaseout,' *Climate Alert* 1: 1.

Anthes, R.A., Cahir, J.J., Fraser, A.B., and Panofsky, H.A. (1980) *The Atmosphere* (3rd edition), Columbus, Ohio: Merrill.

Arthur, L.M. (1988) *The Implication of Climate Change for Agriculture in the Prairie Provinces, CCD 88–01*, Ottawa: Atmospheric Environment Service.

Austin, J. (1983) 'Krakatoa sunsets', *Weather* 38: 226–31.

Bach, W. (1972) *Atmospheric Pollution*, New York: McGraw-Hill.

Bach, W. (1976) 'Global air pollution and climatic change', *Reviews of Geophysics and Space Physics* 14: 429–74.

Bach, W. (1979) 'Short-term climatic alterations caused by human activities: status and outlook', *Progress in Physical Geography* 3: 55–83.

Bach, W. (1986) 'Nuclear war: the effects of smoke and dust on weather and climate', *Progress in Physical Geography* 10: 315–63.

Baker, J.P. and Schofield, C.L. (1985) 'Acidification impacts on fish populations: a review', in D.D. Adams and W.P. Page (eds) *Acid Deposition: Environmental, Economic and Political Issues*, New York: Plenum Press.

Balchin, W.G.V. (1964) 'Hydrology', in J.W. Watson and J.B. Sissons (eds) *The British Isles: A Systematic Geography*, London: Nelson.

Ball, T. (1986) 'Historical evidence and climatic implications of a shift in the boreal forest-tundra transition in Central Canada', *Climatic Change* 8: 121–34.

Bark, L.D. (1978) 'History of American droughts', in N.J. Rosenberg (ed.) *North American Droughts*, Boulder: Westview Press.

Barry, R.G. and Chorley, R.J. (1987) *Atmosphere, Weather and Climate*, London: Methuen.

Berkofsky, L. (1986) 'Weather modification in arid regions: the Israeli experience', *Climatic Change* 9: 103–12.

Biswas, A.K. (1974) *Energy and the Environment*, Ottawa: Environment Canada.

Bolin, B. (1960) 'On the exchange of carbon dioxide between the atmosphere and the sea', *Tellus* 12: 274–81.

Bolin, B. (1972) 'Atmospheric chemistry and environmental pollution', in D.P. McIntyre (ed.) *Meteorological Challenges: A History*, Ottawa: Information Canada.

Bolin, B. (1986) 'How much CO_2 will remain in the atmosphere?', in B. Bolin, B.R. Doos, J. Jager, and R.A. Warrick (eds), *The Greenhouse Effect, Climatic Change and Ecosystems, SCOPE 29*, New York: Wiley.

Bolin, B., Jager, J., and Doos, R.B. (1986) 'The greenhouse effect, climatic change, and ecosystems: a synthesis of present knowledge', in B. Bolin, B.R. Doos, J. Jager, and R.A. Warrick (eds), *The Greenhouse Effect, Climatic Change and Ecosystems, SCOPE 29*, New York: Wiley.

Bolle, H.J., Seiler, W., and Bolin, B. (1986) 'Other greenhouse gases and aerosols', in B. Bolin, B.R. Doos, J. Jager, and R.A. Warrick (eds), *The Greenhouse Effect, Climatic Change and Ecosystems, SCOPE 29*, New York: Wiley.

Borchert, J.R. (1950) 'The climate of the central North American grassland,' *Annals of the Association of American Geographers* 40: 1–39

Bowman, K.P. (1986) 'Interannual variability of total ozone during the breakdown of the Antarctic circumpolar vortex', *Geophysical Research Letters* 13: 1193–6.

Brakke, D.F., Landers, D.H., and Eilers, J.M. (1988) 'Chemical and physical characteristics of lakes in the northeastern United States', *Environmental Science and Technology* 22: 155–63.

Bryson, R.A. (1968) 'All other factors being constant . . .; a reconciliation of several theories of climatic change', *Weather* 21: 56–61 and 94.

Bryson, R.A. (1973) 'Drought in the Sahel: who or what is to blame?', *Ecologist* 3: 366–71.

Bryson, R.A. and Dittberner, G.J. (1976) 'A non-equilibrium model of hemispheric mean surface temperature', *Journal of Atmospheric Sciences* 33: 2094–2106.

Bryson, R.A. and Murray, T.J. (1977) *Climates of Hunger*, Madison: University of Wisconsin Press.

Bryson, R.A. and Peterson, J.T. (1968) 'Atmospheric aerosols: increased concentrations during the past decades', *Science* 162: 120–1.

Burdett, N.A., Cooper, J.R.P., Dearnley, S., Kyte, W.S., and Turnicliffe, M.F. (1985) 'The application of direct limestone injection to U.K. power stations', *Journal of the Institute of Engineering* 58: 64–9.

Burroughs, W.J. (1981) 'Mount St. Helens: a review', *Weather* 36: 238–40.

Calder, N. (1974) *The Weather Machine and the Threat of Ice*, London: BBC Publications.

Callender, G.S. (1938) 'The artificial production of carbon dioxide and its influence on temperature', *Quarterly Journal of the Royal Meteorological Society* 64: 223–37.

Callis, L.B. and Natarajan, M. (1986) 'Ozone and nitrogen dioxide changes in the stratosphere during 1979–1984', *Nature* 323: 772–7.

Canadian International Development Agency (CIDA) (1985) *Food Crisis in Africa*, Hull, Quebec: CIDA.

Catchpole, A.J.W. and Milton, D. (1976) 'Sunnier prairie cities – a benefit of natural gas', *Weather* 31: 348–54.

Caulfield, C. and Pearce, F. (1984) 'Ministers reject clean-up of acid rain', *New Scientist* 104 (1433): 6.

Central Electricity Generating Board (1984) *Acid Rain*, London: CEGB.

Cess, R.D. (1985) 'Nuclear war: illustrative effects of atmospheric smoke and dust upon solar radiation', *Climatic Change* 7: 238–51.

Cess, R.D. and Potter, G.L. (1984) 'A commentary on the recent CO_2-climate controversy', *Climatic Change* 6: 365–76.

Chapman, S. (1930) 'A theory of upper atmospheric ozone', *Quarterly Journal of the Royal Meteorological Society* 3: 103.

Charney, J. (1975) 'Dynamics of deserts and drought in the Sahel', *Quarterly Journal of the Royal Meteorological Society* 101: 193–202.

Chase, A. (1988) 'The ozone precedent', *Outside* 13: 37–40.

Cheng, H.C., Steinberg, M., and Beller, M. (1986) *Effects of Energy Technology on Global CO_2 Emissions*, Washington, DC: US Department of Energy.

Chester, D.K. (1988) 'Volcanoes and climate: recent volcanological perspectives', *Progress in Physical Geography* 12: 1–35.

Cho, H.R., Iribarne, J.V., Grabenstetter, J.E., and Tam, Y.T. (1984) 'Effects of cumulus cloud systems on the vertical distribution of air pollutants', in P.J. Samson (ed.) *The Meteorology of Acid Deposition*, Pittsburgh: Air Pollution Control Association.

Cicerone, R.J., Stolarski, R.S., and Walters, S. (1974) 'Stratospheric ozone destruction by man-made chlorofluoromethanes', *Science* 185: 1165–7.

Climate Institute (1988a) 'NASA scientist testifies greenhouse warming has begun', *Climate Alert* 1: 1–2.

Climate Institute (1988b) 'Is there a drought/greenhouse effect connection', *Climate Alert* 9.

Cortese, T. (1986) 'The Scientific and Economic Implications for Acid Rain Control', in *Conference Proceedings: Intergovernmental Conference on Acid Rain, Quebec, April 10, 11 and 12, 1985*, Quebec City: Ministry of the Environment.

Courel, M.F., Kandel, R.S., and Rasool, S.I. (1984) 'Surface albedo and the Sahel drought', *Nature* 307: 528–31.

Covey, C., Schneider, S.H., and Thompson, S.L. (1984) 'Global effects of massive smoke injections from a nuclear war: results from general circulation model simulations', *Nature* 308: 21–5.

Crane, A. and Liss, P. (1985) 'Carbon dioxide, climate and the sea', *New Scientist* 108: 50–4.

Crawford, M. (1985) 'Sub-Sahara needs quick help to avert disaster', *Science* 230: 788.

Critchfield, H.J. (1983) *General Climatology*, Englewood Cliffs: Prentice-Hall.

Cronan, C.S. and Schofield, C.L. (1979) 'Aluminum leaching response to acid precipitation: effects on high elevation watersheds in the northeast', *Science* 204: 304–6.

Cronin, J.F. (1971) 'Recent volcanism and the stratosphere', *Science* 172: 847–9.

Cross, M. (1985a) 'Africa's drought may last many years', *New Scientist* 105: 9.

Cross, M. (1985b) 'Waiting, and hoping, for the big rains', *New Scientist* 105: 3–4.

Crutzen, P.J. (1972) 'SSTs – A threat to the earth's ozone shield', *Ambio* 1: 41–51.

Crutzen, P.J. (1974) 'Estimates of possible variations in total ozone due to natural causes and human activities', *Ambio* 3: 201–10.

Crutzen, P.J. and Arnold, F. (1986) 'Nitric acid cloud formation in the cold Antarctic stratosphere: a major cause for the springtime "ozone hole"', *Nature* 324: 651–5.

Crutzen, P.J. and Birks, J.W. (1982) 'The atmosphere after a nuclear war: twilight at noon', *Ambio* 11: 114–25.

Crutzen, P.J., Aselmann, I., and Seiler, W. (1986) 'Methane production by domestic animals, wild ruminants, other herbivorous fauna and humans', *Tellus* 38: 271–84.

Crutzen, P.J., Galbally, I.E., and Bruhl, C. (1984) 'Atmospheric effects from post-nuclear fires', *Climatic Change*, 6: 323–64.

Delmas, R., Ascendio, J.M., and Legrand, M. (1980) 'Polar ice evidence that atmospheric CO_2 20,000 yr B.P. was 50% of present', *Nature* 284: 155–7.

Detwyler, T.R. (ed.) (1971) *Man's Impact on Environment*, New York: McGraw-Hill.

Dickinson, R.E. (1986) 'How will climate change', in B. Bolin, B.R. Doos, J. Jager, and R.A. Warrick (eds), *The Greenhouse Effect, Climatic Change and Ecosystems, SCOPE 29*, New York: Wiley.

Dotto, L. and Schiff, H. (1968) *The Ozone War*, Garden City: Doubleday.

Dyer, A.J. and Hicks, B.B. (1965) 'Stratospheric transport of volcanic dust inferred from solar radiation measurements', *Nature* 94: 545–54.

Eagleman, J.R. (1985) *Meteorology: The Atmosphere in Action*, Belmont: Wadsworth.

Edmonds, J.A., Reilly, J.M., Gardner, R.H., and Brenkert, A. (1986) *Uncertainty in Future Global Energy Use and Fossil Fuel CO_2 Emission: 1975 to 2075*, Washington, DC: US Department of Energy.

Ehrlich, P.R. *et al.* (1983) 'Long-term biological consequences of nuclear war', *Science*, 222: 1293–1300.

Ehrlich, P.R., Ehrlich, A.H., and Holdren, J.P. (1977) *Ecoscience: Population, Resources, Environment*, San Francisco: Freeman.

Ellis, E.C., Zeldin, M.D., Brewer, R.L., and Shepard, L.S. (1984) 'Chemistry of Winter Type Precipitation in Southern California', in P.J.

Samson (ed.) *The Meteorology of Acid Deposition*, Pittsburgh: Air Pollution Control Association.

Ellis, H.T. and Pueschel, R.F. (1971) 'Solar radiation: absence of air pollution trends at Mauna Loa', *Science* 172: 845–6.

Enhalt, D.H. (1980) 'The effects of chlorofluoromethanes on climate', in W. Bach, J. Pankrath, and J. Williams (eds) *Interactions of Energy and Climate*, Dordrecht: Reidel.

Environment Canada (1986) *Understanding CO_2 and Climate – Annual Report 1985*, Ottawa: Atmospheric Environment Service.

Environment Canada (1987) *Understanding CO_2 and Climate – Annual Report 1986*, Ottawa: Atmospheric Environment Service.

Evans, W.F.J., McConnell, J.C., Nicholls, R.W., and Vipputuri, R. (1985) 'Some comments on atmospheric modelling in the context of environmental consequences of major nuclear warfare', in The Royal Society of Canada, *Nuclear Winter and Associated Effects*, Ottawa: The Royal Society of Canada.

Farman, J.C., Gardiner, B.G., and Shanklin, J.D. (1985) 'Large losses of total ozone in Antarctica reveal seasonal ClO_x/NO_x interaction', *Nature* 315: 207–10.

Felch, R.E. (1978) 'Drought: characteristics and assessment,' in N.J. Rosenberg (ed.) *North American Droughts*, Boulder, Colorado: Westview Press.

Fennelly, P.F. (1981) 'The origin and influence of airborne particulates', in B.J. Skinner (ed.) *Climates, Past and Present*, Los Altos: Kauffmann.

Fernandez, I.J. (1985) 'Acid deposition and forest soils: potential impacts and sensitivity', in D.D. Adams and W.P. Page (eds) *Acid Deposition: Environmental, Economic and Political Issues*, New York: Plenum Press.

Findley, R. (1981) 'The day the sky fell', *National Geographic* 159: 50–65.

Fischer, W.H. (1967) 'Some atmospheric turbidity measurements in Antarctica', *Journal of Applied Meteorology* 6: 958–9.

Flohn, H. (1980) *Possible Climatic Consequences of a Man-Made Global Warming RR-80-30*, Laxenburg, Austria: International Institute for Applied Systems Analysis.

Foley, H.M. and Ruderman, M.A. (1973) 'Stratospheric NO production from past nuclear explosions', *Journal of Geophysical Research* 78: 4441–50.

Folland, C.K., Palmer, T.N., and Parker, D.E. (1986) 'Sahel rainfall and worldwide sea temperatures 1901–85: observational, modelling, and simulation studies', *Nature* 320: 602–7.

Forster, B.A. (1985) 'Economic impact of acid deposition in the Canadian aquatic sector', in D.D. Adams and W.P. Page (eds) *Acid Deposition: Environmental, Economic and Political Issues*, New York: Plenum Press.

Fowler, M.M. and Barr, S. (1984) 'A long range atmospheric tracer field test', in P.J. Samson (ed.) *The Meteorology of Acid Deposition*, Pittsburgh: Air Pollution Control Association.

Fritts, H. (1965) 'Tree-ring evidence for climatic changes in western North America, *Monthly Weather Review* 93: 421–43.

Garnett, A. (1967) 'Some climatological problems in urban geography with reference to air pollution', *Transactions of the Institute of British Geographers* 42: 21–43.

Glantz, M.H. (ed.) (1977) *Desertification: Environmental Degradation in and around Arid Lands*, Boulder, Colorado: Westview Press.

Goldman, M.I. (1971) 'Environmental disruption in the Soviet Union', in T.R. Detwyler (ed.) *Man's Impact on Environment*, New York: McGraw-Hill.

Greene, O., Percival, I., and Ridge, I. (1985) *Nuclear Winter: The Evidence and The Risks*, Cambridge: Polity Press.

Gribbin, J. (1978) 'Fossil fuel: future shock?', *New Scientist* 79: 541–3.

Gribbin, J. (1981) 'The politics of carbon dioxide', *New Scientist* 90: 82–4.

Gribbin, J. (1982) *Future Weather and the Greenhouse Effect*, New York: Delacorte.

Grover, H.D. (1985) 'The biological consequences of nuclear war for Canada', in The Royal Society of Canada *Nuclear Winter and Associated Effects*, Ottawa: The Royal Society of Canada.

Hainsworth, P.H. (1985) 'Nuclear winter,' letter in *Weather* 40: 62.

Hamilton, R.A. and Archbold, J.W. (1945) 'Meteorology over Nigeria and adjacent territory', *Quarterly Journal of the Royal Meteorological Society* 71: 231–62.

Hammond, A.L. and Maugh, T.H. (1974) 'Stratospheric pollution: multiple threats to earth's ozone', *Science* 186: 335–8.

Hampson, J. (1974) 'Photochemical war on the atmosphere', *Nature* 250: 189–91.

Hansen, J. and Lebedeff, S. (1988) 'Global surface air temperatures: update through 1987' *Geophysical Research Letters* 15(4): 323–6.

Hare, F.K. (1973) 'The atmospheric environment of the Canadian north', in D.H. Pimlott, K.M. Vincent, and C.E. McKnight (eds) *Arctic Alternatives*, Ottawa: Canadian Arctic Resources Committee.

Hare, F.K. and Thomas, M.K. (1979) *Climate Canada*, Toronto: John Wiley.

Harwell, M.A. and Hutchinson, T.C. (1985) *Environmental Consequences of Nuclear War, Volume II: Ecological and Agricultural Effects*, SCOPE 28, New York: Wiley.

Hekstra, G.P. (1986) 'Will climatic changes flood the Netherlands? Effects on agriculture, land-use and well-being', *Ambio* 15: 316–26.

Hekstra, G.P. and Liverman, D.M. (1986) 'Global food futures and desertification', *Climatic Change* 9: 59–66.

Henderson-Sellers, A. (1986) 'Cloud changes in a warmer Europe', *Climatic Change* 8: 25–52.

Hendrey, G.R. (1985) 'Acid deposition: a national problem', in D.D. Adams and W.P. Page (eds) *Acid Deposition: Environmental, Economic and Political Issues*, New York: Plenum Press.

Henriksen, A. and Brakke, D.F. (1988) 'Sulphate deposition to surface water', *Environmental Science and Technology* 22: 8–14.

Hepting, G.H. (1971) 'Damage to forest from air pollution', in T.R. Detwyler (ed.) *Man's Impact on Environment*, New York: McGraw-Hill.

Hoffman, J.S., Keyes, D., and Titus, J.G. (1983) *Projecting Future Sea Level Rise*, US Environmental Protection Agency: Washington, DC.

Idso, S.B. (1980) 'The climatological significance of doubling of Earth's atmospheric carbon dioxide concentration', *Science*, 207: 1462–3.

Idso, S.B. (1981) 'Carbon dioxide – an alternative view', *New Scientist* 92: 444–6.

Idso, S.B. (1982) *Carbon Dioxide: Friend or Foe*, Tempe, Arizona: IBR Press.

Idso, S.B. (1983) 'Do increases in atmospheric CO_2 have a cooling effect on surface air temperature?', *Climatological Bulletin* 17: 22–6.

Idso, S.B. (1987) 'A clarification of my position on the CO_2/climate connection', *Climatic Change* 10: 81–6.

Israelson, D. (1987) 'Acid rain update', *The Toronto Star* 4 April, A1–A3.

Jenkins, I. (1969) 'Increases in averages of sunshine in Greater London', *Weather* 24: 52–4.

Jensen, K.W. and Snekvik, E. (1972) 'Low pH levels wipe out salmon and trout populations in southernmost Norway', *Ambio* 1: 223–5.

Johnson, A.H. and Siccama, T.G. (1983) 'Acid deposition and forest decline', *Environmental Science and Technology* 17: 294–305.

Johnson, F.S. (1972) 'Ozone and SSTs', *Biological Conservation* 4: 220–2.

Johnston, H.S. (1971) 'Reduction of stratospheric ozone by nitrogen oxide catalysts from SST exhaust', *Science* 173: 517–22.

Kamara, S.I. (1986) 'The origins and types of rainfall in West Africa', *Weather* 41: 48–56.

Katz, R.W. and Glantz, M.H. (1977) 'Rainfall statistics, droughts and desertification in the Sahel', in M.H. Glantz (ed.) *Desertification: Environmental Degradation in and around Arid Lands*, Boulder, Colorado: Westview Press.

Keepin, W., Mintzer, I., and Kristoferson, L. (1986) 'Emission of CO_2 into the atmosphere', in B. Bolin, B.R. Doos, J. Jager, and R.A. Warrick (eds) *The Greenhouse Effect, Climatic Change and Ecosystems, SCOPE 29*, New York: Wiley.

Kellogg, W.W. (1980) 'Aerosols and climate', in W. Bach, J. Pankrath, and J. Williams (eds) *Interactions of Energy and Climate*, Dordrecht: Reidel.

Kellogg, W.W. (1987) 'Mankind's impact on climate: the evolution of an awareness', *Climatic Change* 10: 113–36.

Kellogg, W.W. and Schneider, S.H. (1977), 'Climate, desertification and human activities', in M.H. Glantz (ed.) *Desertification: Environmental Degradation in and around Arid Lands*, Boulder, Colorado: Westview Press.

Kleinbach, M.H. and Salvagin, C.E. (1986) *Energy Technologies and Conversion Systems*, Englewood Cliffs: Prentice Hall.

Komhyr, W.D., Grass, R.D., and Leonard, R.K. (1986) 'Total ozone decrease at the South Pole 1964–1985, *Geophysical Research Letters* 13: 1248–51.

Kyte, W.S. (1981) 'Some chemical and chemical engineering aspects of flue gas desulphurization', *Transactions of the Institute of Chemical Engineers* 59: 219–29.

Kyte, W.S. (1986a) 'Some aspects of possible control technologies for coal-fired power stations in the United Kingdom', *Mine and Quarry* 15: 26–9.

Kyte, W.S. (1986b) 'Possible emissions control technologies for coal-fired power stations – the United Kingdom', paper presented at the *Institution of Chemical Engineers Conference on 'The Problem of Acid Emission'*, University of Birmingham.

Kyte, W.S. (1988) 'A programme for reducing SO_2 emissions from U.K. power stations – present and future', *CEGB: Corporate Environment Unit*, Paper no. 1–2.

LaBastille, A. (1981) 'Acid rain – how great a menace?', *National Geographic* 160: 652–81.

Lamb, H.H. (1970) 'Volcanic dust in the atmosphere; with a chronology and assessment of its meteorological significance', *Philosophical Transactions of the Royal Society, A* 266: 435–533.

Lamb, H.H. (1972) *Climate: Present, Past and Future: Volume 1. Fundamentals and Climate Now*, London: Methuen.

Lamb, H.H. (1977) *Climate: Present, Past and Future: Volume 2, Climatic History and the Future*, London: Methuen.

Landsberg, H.H. (1986) 'Potentialities and limitations of conventional climatological data for desertification monitoring and control', *Climatic Change* 9: 123–8.

Lane, P. and Associates Ltd (1988) *Preliminary Study of the Possible Impacts of a One Metre Rise in Sea Level at Charlottetown, Prince Edward Island CCD88-02*, Ottawa: Atmospheric Environment Service.

Last, F.T. and Nicholson, I.A. (1982) 'Acid rain', *Biologist* 29: 250–2.

Laval, K. (1986) 'General circulation model experiments with surface albedo changes, *Climatic Change* 9: 91–102.

Lawson, M.P. and Stockton, C. (1981) 'The desert myth evaluated in the context of climatic change', in C.D. Smith and M. Parry (eds) *Consequences of Climatic Change*, Nottingham: University of Nottingham.

Le Houérou, H.N. (1977) 'The nature and causes of desertization', in M.H. Glantz (ed.) *Desertification: Environmental Degradation in and around Arid Lands*, Boulder, Colorado: Westview Press.

Leighton, P.A. (1966) 'Geographical aspects of air pollution', *Geographical Review* 56: 151–74.

Lemonick, M.D. (1987) 'The heat is on', *Time* 130 (16): 62–75.

Lippmann, M. (1986) 'The effects of inhaled acid on human health', in *Conference Proceedings: Intergovernmental Conference on Acid Rain, Quebec, April 10, 11 and 12, 1985*, Quebec City: Ministry of the Environment.

Lockwood, J.G. (1979) *Causes of Climate*, London: Edward Arnold.

Lockwood, J.G. (1984) 'The southern oscillation and El Niño', *Progress in Physical Geography* 8: 102–10.

Lockwood, J.G. (1986) 'The causes of drought with particular reference to the Sahel', *Progress in Physical Geography* 10: 110–19.

Lovelock, J.E. (1972) 'Gaia as seen through the atmosphere', *Atmospheric Environment* 6: 579–80.

Lovelock, J.E. (1986) 'Gaia: the world as living organism', *New Scientist* 112: 25–8.

Lovelock, J.E. and Epton, S. (1975) 'The quest for Gaia', *New Scientist* 65: 304–6.

Lovelock, J.E. and Margulis, L. (1973) 'Atmospheric homeostasis by and for the biosphere: the Gaia hypothesis', *Tellus* 26: 2.

Ludlum, D.M. (1971) *Weather Record Book*, Princeton, New Jersey: Weatherwise Inc.

Lutgens, F.K. and Tarbuck, E.J. (1982) *The Atmosphere*, Englewood Cliffs: Prentice-Hall.

McCarthy, J.J., Brewer, P.G., and Feldman, G. (1986) 'Global ocean flux', *Oceanus* 29: 16–26.

McCormick, R.A. and Ludwig, J.L. (1976) 'Climate modification by atmospheric aerosols', *Science* 156: 1358–9.

MacCracken, M. (1983) 'Nuclear war: preliminary estimates of the climatic effects of nuclear exchange', *Proceedings of the Third International Seminar on Nuclear War*, Erice, Sicily: 161–83.

MacCracken, M.C. and Luther, F.M. (1985a) *Detecting the Climatic Effects of Increasing Carbon Dioxide*, Washington, DC: US Department of Energy.

MacCracken, M.C. and Luther, F.M. (1985b) *Projecting the Climatic Effects of Increasing Carbon Dioxide*, Washington, DC: US Department of Energy.

McKay, G.A., Maybank, J., Mooney, O.R., and Pelton, W.L. (1967) *The Agricultural Climate of Saskatchewan: Climatological Studies, no. 10*, Toronto: Canada, Department of Transport, Meteorological Branch.

Mackenzie, D. (1987a) 'Ethiopia plunges towards another famine', *New Scientist* 115: 26.

Mackenzie, D. (1987b) 'Can Ethiopia be saved?' *New Scientist* 115: 54–8.

Maddox, J. (1984) 'Nuclear winter not yet established', *Nature* 308: 11

Maddox, J. (1988) 'What happened to nuclear winter?', *Nature* 333: 203.

Manabe, S. and Wetherald, R.T. (1975) 'The effects of doubling the CO_2 concentration on the climate of a general circulation model', *Journal of Atmospheric Science* 32: 3–15.

Manabe, S. and Wetherald, R.T. (1986) 'Reduction in summer soil wetness induced by an increase in atmospheric carbon dioxide', *Science* 232: 626–8.

Manabe, S., Wetherald, R.T., and Stouffer, R.J. (1981) 'Summer dryness due to an increase of atmospheric CO_2 concentration', *Climatic Change* 3: 347–86.

Manson, A.N. (1985) 'Acid deposition: the Canadian perspective' in D.D. Adams and W.P. Page (eds) *Acid Deposition: Environmental,*

Economic and Political Issues, New York: Plenum Press.

Martec Ltd (1987) *Effects of a One metre Rise in Sea Level at St. John, New Brunswick and the Lower Reaches of the St. John River CCD 87-04*, Ottawa: Atmospheric Environment Service.

Meinl, H., Bach, W., Jager, J., Jung, H.J., Knottenberg, H., Marr, G., Santer, B.D., and Schwieren, G. (1984) *The Socio-Economic Impacts of Climatic Changes due to a Doubling of Atmospheric CO_2 Content*, Brussels: Commission of European Communities Report.

Miller, J.M. (1984) 'Acid rain', *Weatherwise* 37: 234-9.

Mitchell, J.F.B. (1983) 'The seasonal response of a general circulation model to changes in CO_2 and sea temperature', *Quarterly Journal of the Royal Meteorological Society*, 109: 113-52.

Mitchell, J.M. (1975) 'A reassessment of atmospheric pollution as a cause of long-term changes of global temperature', in S.F. Singer (ed.) *The Changing Global Environment*, Boston: Reidel.

Molina, M.J. and Rowland, F.S. (1974) 'Stratospheric sink for chlorofluoromethanes: chlorine atom-catalysed destruction of ozone', *Nature* 249: 810-2.

Moore, B. and Bolin, B. (1986) 'The oceans, carbon dioxide and global climate change', *Oceanus*, 29: 9-15.

Morales, C. (1986) 'The Airborne Transport of Saharan Dust: A Review', *Climatic Change* 9: 219-42.

Mossop, S.C. (1964) 'Volcanic dust collected at an altitude of 20 km', *Nature* 203: 824-7.

Murphey, R. (1973) *The Scope of Geography*, Chicago: Rand McNally.

Musk, L. (1983) 'Outlook-changeable', *Geographical Magazine*, 55: 532-3.

National Research Council (NRC) (1982) *Carbon Dioxide and Climate: A Second Assessment*, Washington, DC: National Academy Press.

National Research Council (NRC) (1983) *Changing Climate*, Washington, DC: National Academy Press.

National Research Council (NRC) (1985) *The Effects on the Atmosphere of a Major Nuclear Exchange*, Washington, DC: National Academy Press.

Newell, R.E. and Dopplick, T.G. (1979) 'Questions concerning the possible influence of anthropogenic CO_2 on atmospheric temperature', *Journal of Applied Meteorology* 18: 822-5.

Newhall, G.G. and Self, S. (1982) 'The volcanic explosivity index (VEI): an estimate of the explosive magnitude for historical volcanism', *Journal of Geophysical Research* 87: 1231-8.

Nicholson, S.E. (1989) 'Long-term changes in African rainfall', *Weather* 44: 47-56.

Norton, P. (1985) 'Decline and fall', *Harrowsmith* 9: 24-43.

Oguntoyinbo, J. (1986) 'Drought prediction', *Climatic Change* 9: 79-90.

Ontario: Ministry of the Environment (1980) *The Case Against the Rain*, Toronto: Information Services Branch, Ministry of the Environment.

Owen, J.A. and Ward, M.N. (1989) 'Forecasting Sahel rainfall', *Weather*, 44: 57-64.

Park, C.C. (1987) *Acid Rain: Rhetoric and Reality*, London: Methuen.

Parry, M.L. (1978) *Climatic Change, Agriculture and Settlement*, Folke-stone: Dawson.

Pearce, F. (1982a) 'Warning cones hoisted as acid rain-clouds gather', *New Scientist* 94: 828.

Pearce, F. (1982b) 'Science and politics don't mix at acid rain debate', *New Scientist* 95: 3.

Pearce, F. (1982c) 'It's an acid wind that blows nobody any good', *New Scientist* 95: 80.

Pearce, F. (1982d) 'The menace of acid rain', *New Scientist* 95: 419–23.

Pearce, F. (1984) 'MPs back curbs on acid rain', *New Scientist* 103: 3.

Peczkic, J. (1988) 'Initial uncertainties in "Nuclear Winter": a proposed test based on the Dresden firestorm', *Climatic Change* 12: 199–208.

Peterson, J. (ed.) (1983) *Nuclear War: The Aftermath*, Oxford: Pergamon.

Peterson, J.T. and Junge, C.E. (1971) 'Sources of particulate matter in the atmosphere', in W.H. Matthews, W.W. Kellogg, and G.D. Robinson (eds) *Man's Impact on the Climate*, Cambridge, Mass.: MIT Press.

Phillips, D.W. (1982) *Climatic anomalies and unusual weather in Canada during 1981*, Downsview, Ontario: Atmospheric Environment Service.

Piette, J. (1986) 'Les initiatives des provinces de l'Est du Canada en matiere de precipitations acides', in *Conference Proceedings: Inter-governmental Conference on Acid Rain, Quebec, April 10, 11 and 12, 1985*, Quebec City: Ministry of the Environment.

Pisias, N.G. and Imbrie, J. (1986) 'Variations in the earth's orbit influenced past climate changes', *Oceanus* 29: 43–9.

Pittock, A.B. and Salinger, M.J. (1982) 'Towards regional scenarios for a CO_2-warmed earth', *Climatic Change* 4: 23–40.

Pittock, A.B., Ackerman, T.P., Crutzen, P.J., MacCracken, M.C., Shapiro, C.S., and Turco, R.P. (1986) *Environmental Consequences of Nuclear War, Volume 1: Physical and Atmospheric Effects, SCOPE 28*, New York: Wiley.

Pollack, J.B. and Ackerman, T.P. (1983) 'Possible effects of the El Chichon volcanic cloud on the radiation budget of the northern tropics', *Geophysical Research Letters* 10: 1057–60.

Ponte, L. (1976) *The Cooling*, Englewood Cliffs: Prentice-Hall.

Potter, G.L., Kiehl, J.T., and Cess, R.D. (1987) 'A clarification of certain issues related to the CO_2-climate problem', *Climatic Change* 10: 87–95.

Ramage, J. (1983) *Energy: A Guidebook*, Oxford: Oxford University Press.

Ramanathan, V. (1975) 'Greenhouse effect due to chlorofluorocarbons: climatic implications', *Science* 190: 50–1.

Ramanathan, V., Callis, L.B., and Boughner, R.E. (1976) 'Sensitivity of surface temperature and atmospheric temperature to perturbations in the stratospheric concentration of ozone and nitrogen dioxide', *Journal of Atmospheric Sciences* 33: 1092–112.

Ramanathan, V., Singh, H.B., Cicerone, R.J., and Kiehl, J.T. (1985) 'Trace gas trends and their potential role in climate change', *Journal of*

Geophysical Research 90: 5547–56.

Rampino, M.R. and Self, S. (1984) 'The atmospheric effects of El Chichon', *Scientific American* 250: 48–57.

Rasmusson, E.M. and Hall, J.M. (1983) 'El Niño', *Weatherwise* 36(4): 166–75.

Revelle, R. and Seuss, H.E. (1957) 'Carbon dioxide exchange between the atmosphere and ocean and the question of an increase of atmospheric CO_2 during the past decades', *Tellus* 9: 18–27.

Riefler, R.F. (1978) 'Drought: an economic perspective', in N.J. Rosenberg (ed.) *North American Droughts*, Boulder, Colorado: Westview Press.

Robitaille, L. (1986) 'Le déperissement des erablières au Quebec; problematique et état des recherches', in *Conference Proceedings: Intergovernmental Conference on Acid Rain, Quebec, April 10, 11 and 12, 1985*, Quebec City: Ministry of the Environment.

Robock, A. (1984) 'Snow and ice feedbacks prolong effects of nuclear winter', *Nature* 310: 667–70.

Rosenberg, N.J. (1978) *North American Droughts*, Boulder, Colorado: Westview Press.

Rosenthal, H. and Wilson, J.S. (1987) *An Updated Bibliography (1945–1986) on Ozone, its Biological Effects and Technical Applications*, Halifax, Nova Scotia: Ministry of Supply and Services, Canada.

Royal Society of Canada (RSC) (1985) *Nuclear Winter and Associated Effects: A Canadian Appraisal of the Environmental Impact of Nuclear War*, Ottawa: The Royal Society of Canada.

Sagan, C. (1983) 'Nuclear war and climatic catastrophe: some policy implications', *Foreign Affairs* 62: 257–92.

Sage, B. (1980) 'Acid drops from fossil fuels', *New Scientist* 85: 743–5.

Sanders, D.W. (1986) 'Desertification processes and impact in rainfed agricultural regions', *Climatic Change* 9: 33–42.

Sanderson, M. (1987) *Implications of Climatic Change for Navigation and Power Generation in the Great Lakes, CCD 87–03*, Ottawa: Atmospheric Environment Service.

Santer, B. (1985) 'The use of general circulation models in climate impact analysis – a preliminary study of the impact of CO_2-induced climatic change on West European agriculture', *Climatic Change* 7: 71–93.

SCEP (1970) *Man's Impact on the Global Environment: Study of Critical Environmental Problems*, Cambridge, Mass.: MIT Press.

Scheider, W., Adamski, J., and Paylor, M. (1975) *Reclamation of Acidified Lakes near Sudbury, Ontario*, Toronto, Ontario: Ministry of the Environment.

Schneider, S.H. (1978) 'Forecasting future droughts: is it possible?', in N.J. Rosenberg (ed.) *North American Droughts*, Boulder, Colorado: Westview Press.

Schneider, S.H. (1984) 'National experiments and CO_2-induced climate change: the controversy drags on – an editorial', *Climatic Change* 6: 317–21.

Schneider, S.H. (1986) 'A goddess of the earth?: the debate on the Gaia hypothesis – an editorial', *Climatic Change* 8: 1–4.

Schneider, S.H. (1987) 'Climatic modeling', *Scientific American* 256: 72–89.

Schneider, S.H. (1988) 'Whatever happened to nuclear winter? – an editorial', *Climatic Change* 12: 215–9.

Schneider, S.H. and Mass, C. (1975) 'Volcanic dust, sunspots and temperature trends', *Science* 190: 741–6.

Schneider, S.H. and Mesirow, L. (1976) *The Genesis Strategy*, New York: Plenum Press.

Schneider, S.H. and Thompson, S.L. (1981) 'Atmospheric CO_2 and climate: importance of the transient response', *Journal of Geophysical Research*, 86: 3135–47.

Schneider, S.H. and Thompson, S.L. (1988) 'Simulating the climatic effects of nuclear war', *Nature* 333: 221–7.

Schoeberl, M.R. and Krueger, A.J. (1986) 'Overview of the Antarctic ozone depletion issue', *Geophysical Research Letters* 13: 1191–2.

Sellers, W.D. (1973) 'A new global climatic model', *Journal of Applied Meteorology* 12: 241–54.

Shaw, G.E. (1980) 'Arctic Haze', *Weatherwise* 33: 219–21.

Shaw, R.W. (1987) 'Air pollution by particles', *Scientific American* 257: 96–103.

Shine, K. (1988) 'Antarctic ozone – an extended meeting report', *Weather* 43: 208–10.

Shriner, D.S. and Johnston, J.W. (1985) 'Acid rain interactions with leaf surfaces: a review', in D.D. Adams and W.P. Page (eds) *Acid Deposition: Environmental, Economic and Political Issues*, New York: Plenum Press.

Shugart, H.H., Antonovsky, P.G., Jarvis, P.G., and Sandford, A.P. (1986) 'CO_2, climatic change and forest ecosystems', in B. Bolin, B.R. Doos, J. Jager, and R.A. Warrick (eds), *Greenhouse Effect, Climatic Change and Ecosystems, SCOPE 29*, New York: Wiley.

Siegenthaler, U. (1984) '19th century measurements of atmospheric CO_2 – a comment', *Climatic Change* 6: 409–11.

Singer, S.F. (1984) 'Is the "nuclear winter" real?', *Nature* 310: 625.

SMIC (1971) *Inadvertent Climate Modification: Report of the Study of Man's Impact on Climate*, Cambridge, Mass.: MIT Press.

Smit, B. (1987) *Implications of Climatic Change for Agriculture in Ontario CCD 87–02*, Ottawa: Atmospheric Environment Service.

Smith, J.W. (1920) 'Rainfall of the Great Plains in relation to cultivation', *Annals of the Association of American Geographers* 10: 69–74.

Spry, I.M. (1963) *The Palliser Expedition*, Toronto: MacMillan.

Starr, V.P. (1956) 'The general circulation of the atmosphere', *Scientific American* 195: 40–5.

Steila, D. (1976) *The Geography of Soils*, Englewood Cliffs: Prentice-Hall.

Stokoe, P. (1988) *Socio-economic Assessment of the Physical and Ecological Impacts of Climate Change on the Marine Environment of the Atlantic Region of Canada – Phase 1 CCD 88–07*, Ottawa: Atmospheric Environment Service.

Stolarski, R.S. and Cicerone, R.J. (1974) 'Stratospheric chlorine. Possible sink for ozone', *Canadian Journal of Chemistry* 52: 1610–5.

Stolarksi, R.S., Krueger, A.J., Schoeberl, M.R., McPeters, R.D., Newman, P.A., and Alpert, J.C. (1986) 'Nimbus 7 satellite measurements of the springtime Antarctic ozone decrease', *Nature* 322: 808–11.

Strain, B.R. and Cure, J.D. (eds) (1985) *Direct Effects of Increasing Carbon Dioxide on Vegetation*, Washington, DC: US Department of Energy.

Stuiver, M. (1978) 'Atmospheric carbon dioxide and carbon reservoir changes', *Science*, 199: 253–8.

Sveinbjornsson, B. (1984) 'Alaskan plants and atmospheric carbon dioxide', in J.H. McBeath (ed.), *The Potential Effects of Carbon Dioxide, Conference Proceedings*, Fairbanks: School of Agriculture and Land Resources Management, University of Alaska.

Sweeney, J. (1985) 'The 1984 Drought on the Canadian Prairies', *Weather* 40: 302–9.

Swift, J. (1977) 'Pastoral development in Somalia: herding cooperatives as a strategy against desertification and famine', in M. Glantz (ed.) *Desertification: Environmental Degradation in and around Arid Lands*, Boulder, Colorado: Westview Press.

Teller, E. (1984) 'Widespread after-effects of nuclear war', *Nature*, 310: 621–4.

Thackrey, T.O. (1971) 'Pittsburgh: how one city did it', in R. Revelle, A. Khosla, and M. Vinovskis (eds) *The Survival Equation*, Boston: Houghton Mifflin.

Thompson, S.L. (1984) 'An evolving "nuclear winter" – guest editorial', *Climatic Change*, 6: 105–7.

Thompson, S.L. (1985) 'Global interactive transport simulations of nuclear war smoke', *Nature* 317: 35–9.

Thompson, S.L. and Schneider, S.H. (1986) 'Nuclear winter reappraised', *Foreign Affairs* 64: 981–1005.

Thornthwaite, C.W. (1947) 'Climate and moisture conservation', *Annals of the Association of American Geographers* 37: 87–100.

Titus, J.G. (1986) 'Greenhouse effect, sea level rise and coastal zone management', *Coastal Zone Management Journal* 14: 147–72.

Tomlinson, G.H. (1985) 'Forest vulnerability and the cumulative effects of acid deposition', in D.D. Adams and W.P. Page (eds) *Acid Deposition: Environmental Economic and Political Issues*, New York: Plenum Press.

Toon, O.B. and Pollack, J.B. (1981) 'Atmospheric aerosols and climate', in B.J. Skinner (ed.) *Climates Past and Present*, Los Altos: Kauffmann.

Trewartha, G.T. and Horn, L.H. (1980) *An Introduction to Climate*, New York: McGraw-Hill.

Tucker, A. (1987) 'Ozone hole culprit pin-pointed', *Manchester Guardian Weekly* 137(8): 4.

Tullett, M.T. (1984) 'Saharan dust-fall in Northern Ireland', *Weather* 39: 151–2.

Turco, R.P. and Ackerman, T.P. (1985) 'A Canadian nuclear winter', in

The Royal Society of Canada, *Nuclear Winter and Associated Effects*, Ottawa: The Royal Society of Canada.

Turco, R.P., Toon, O.B., Ackerman, T.P., Pollack, J.B., and Sagan, C. (1983) 'Nuclear winter: global consequences of multiple nuclear explosions', *Science* 222: 1283–92.

Turk, J. and Turk, A. (1988) *Environmental Science*, Philadelphia: Saunders.

Valerie, K., Delers, A., Bruck, C., Thiriart, C., Rosenberg, H., Debouck, C., and Rosenberg, M. (1988) 'Activation of human immunodeficiency virus type 1 by DNA damage in human cells', *Nature* 333: 78–81.

Van Royen, W. (1937) 'Prehistoric droughts in the central Great Plains', *Geographical Review* 27: 637–50.

van Ypersele, J.P. and Verstraete, M.M. (1986) 'Climate and desertification – editorial, *Climatic Change* 9: 1–4.

Verstraete, M.M. (1986) 'Defining desertification: a review', *Climatic Change* 9: 5–18.

Viereck, L.A. and Van Cleve, K. (1984) 'Some aspects of vegetation and temperature relationships in the Alaska taiga', in J.H. McBeath (ed.) *The Potential Effects of Carbon Dioxide-Induced Climatic Change in Alaska: Conference Proceedings*, Fairbanks: School of Agriculture and Land Resources Management, University of Alaska.

Wall, G. (1988) *Implications of Climatic Change for Tourism and Recreation in Ontario CCD 88–05*, Ottawa: Atmospheric Environment Service.

Ware, H. (1977) 'Desertification and population: sub-Saharan Africa', in M.H. Glantz (ed.) *Desertification: Environmental Degradation in and around Arid Lands*, Boulder, Colorado: Westview Press.

Washington, W.M. and Meehl, G.A. (1984) 'Seasonal cycle experiment on the climate sensitivity due to a doubling of CO_2 with an atmospheric general circulation model coupled to a simple mixed-layer ocean model', *Journal of Geophysical Research* 89: 9475–503.

Watson, J.W. (1963) *North America: Its Countries and Regions*, London: Longman.

Webster, M. (1988) 'Queries and quandries', *Harrowsmith* 80: 115–16.

Webster, P.J. (1984) 'The carbon dioxide/climate controversy: some personal comments on two recent publications', *Climatic Change* 6: 377–90.

Wigley, T.M.L., Jones, P.D., and Kelly, P.M. (1986) 'Empirical climate studies' in B. Bolin, B.R. Doos, J. Jager, and R.A. Warrick (eds) *The Greenhouse Effect, Climatic Change and Ecosystems, SCOPE 29*, New York: Wiley.

Williams, G.D.V., Fantley, R.A., Jones, K.H., Stewart, R.B., and Wheaton, E.E. (1988) *Estimating Effects of a Climatic Change on Agriculture in Saskatchewan CCD 88–06*, Ottawa: Atmospheric Environment Service.

Williamson, S.J. (1973) *Fundamentals of Air Pollution*, Reading, Mass.: Addison-Wesley.

Wilson, A.T. (1978) 'Pioneer agricultural explosion of CO_2 levels in the atmosphere', *Nature* 273: 40–1.

Wilson, C.A. and Mitchell, J.F.B. (1987) 'Simulated climate and CO_2-induced climate change over western Europe', *Climatic Change* 10: 11–42.

Wittwer, S. (1984) 'The rising level of atmospheric carbon dioxide: an agricultural perspective', in J.H. McBeath (ed.) *The Potential Effects of Carbon Dioxide-Induced Climatic Change in Alaska: Conference Proceedings*, Fairbanks: School of Agriculture and Land Resources Management, University of Alaska.

Wofsy, S.C., McElroy, M.B., and Sze, N.D. (1975) 'Freon consumption: Implications for atmospheric ozone', *Science* 187: 535–7.

Index

215